Calculus

SIMPLIFIED
AND SELF - TAUGHT

SANDERSON M. SMITH
and
FRANK W. GRIFFIN

Mathematics Department
Cate School

ARCO PUBLISHING, INC.
NEW YORK

Dedicated to

Mr. and Mrs. Ephraim K. Smith
Mr. and Mrs. William M. Griffin
Marion McCallum

Published by Arco Publishing, Inc.
215 Park Avenue South, New York, N.Y. 10003

Library of Congress Cataloging in Publication Data
Smith, Sanderson M.
  Calculus simplified and self-taught.

    1. Calculus. I. Griffin, Frank W. II. Title.
QA303.S655 1985   515   84-6466
ISBN 0-668-05756-4 (Paper Edition)

Printed in the United States of America

CONTENTS

INTRODUCTION

As the title suggests, this book is not highly theoretical. Our purpose is to develop within the reader an ability to work with concepts and ideas that one would find in a first-year calculus course. We attempt to show how to use calculus techniques rather than to provide elaborate theoretical details relating to why the techniques are valid. However, while we do not want this book to be a complicated text, we also do not want it merely to be a list of formulas and techniques followed by examples. So, when appropriate, we have provided short explanations as to why a given process is reasonable. We attempted to draw upon our combined twenty-eight years of teaching experience when we inserted what we hope are helpful explanations.

As for content, the book covers the topics listed in the syllabus of the College Entrance Examination Board's Calculus AB Examination. Calculus AB is defined to be a one-year course in elementary functions which places emphasis on an intuitive understanding of the concepts of differential and integral calculus. Some sections of the book are marked with an asterisk (*). These sections contain material that could be omitted by a student whose sole purpose is to study Calculus AB material. These sections contain material that is necessary for the Calculus BC Advanced Placement Examination. Calculus BC typically prepares a student for placement one semester beyond that granted for a corresponding performance in Calculus AB. Most of these sections appear at the end of the book. However, others appear scattered throughout the book because they contain material closely related to the preceding AB material. A section number containing an A (for instance, *Section 1.8A) contains advanced material relating directly to the preceding section (Section 1.8).

The book has been designed for self-study, but it could certainly be used as a classroom textbook for a first year non-theoretical calculus course. Many colleges are instituting more rigid mathematics requirements, including calculus, with emphasis on using calculus techniques which are important in engineering, economics, statistics, and other disciplines. Features of this book include:

(1) Shaded portions in each section. These contain the "meat" of the section. An individual using the book for review could merely read the shaded sections, using the examples, authors' notes, and problems only if he or she determines that more intensive study is needed for that particular topic.

(2) Short sections . . . and to the point. The question and answer format enables readers to learn by doing, rather than by reading about how to do. When we deemed it appropriate, we inserted comments that may be beneficial to the reader.

(3) Authors' Notes, easily identified by the symbol on the right.
Here we are attempting to talk to the reader just as a teacher might talk to his or her students. Basically, we are saying, "Hey, we want to point something out to you here."

(4) Common Errors, easily identified by the symbol on the right.
Here we are attempting to spotlight typical errors that are made with the concepts in the section. This type of analysis should be a great help to those who are using the book to prepare for a test. In general, if you understand why a statement is in error, you should have a good understanding of the topic . . . and, we hope, will not make that type of error on your test.

(5) Questions (*Q1*, *Q2*, etc.) within sections. These are usually short-answer questions to see if the reader "gets the idea" of the section. Readers can attempt to do the problems without looking at the answer (which is in a box below the problem) or they can read the answer as part of the text-reading process.

(6) Each section is followed by a problem set covering the concepts presented in the section. Solutions, answers, and comments are provided for each problem set. In the comment section we are again attempting to "talk to the reader" by pointing out things that should be noted and that may enhance the learning process. The problem sets are not lengthy and, in most cases, do not stress repetition. Rather, each problem has been chosen to highlight a certain concept or idea.

(7) A multiple-choice test appears at the end of each chapter. Its purpose is to reinforce the material within the chapter. A student who can understand and work the multiple-choice problems has a good grasp of the chapter content.

(8) Finally, a set of detailed supplemental problems has been provided at the end of the text. Comments following each of these problems direct the reader to the section to review if he or she experiences difficulty in solving a specific problem.

As should be expected, this book assumes a sound knowledge of algebra, geometry, and trigonometry. Reference and review formulas relating to these branches of mathematics follow.

REFERENCE AND REVIEW FORMULAS

I. Algebra

Quadratic formula: Given $ax^2 + bx + c = 0$, $a \neq 0$, then

$$x = \frac{-b \pm \sqrt{b^2 - 4ac}}{2a}.$$

If $b^2 - 4ac > 0$, the roots are real and unequal.
If $b^2 - 4ac = 0$, the roots are real and equal.
If $b^2 - 4ac < 0$, the roots are complex and unequal.

Binomial Theorem: $(a + b)^n = a^n + na^{n-1}b + \dfrac{n(n-1)}{1 \cdot 2}a^{n-2}b^2 + \cdots + b^n.$

Exponents: $\quad a^x a^y = a^{x+y} \qquad\qquad (a^x)^y = a^{xy} \qquad\qquad (ab)^x = a^x b^x$

$$a^{-x} = \frac{1}{a^x} \qquad\qquad a^0 = 1 \quad (a \neq 0)$$

Logarithms: $\quad y = \log_b a \Leftrightarrow b^y = a \quad (a > 0, b > 0, b \neq 1)$

$$\log_b(ac) = \log_b a + \log_b c \qquad\qquad \log_b\left(\frac{a}{c}\right) = \log_b a - \log_b c$$

$$\log_b a^x = x(\log_b a) \qquad\qquad \log_b \sqrt[n]{a} = \frac{1}{n}\log_b a$$

$$\log_b a = \frac{\log_c a}{\log_c b} \qquad\qquad \ln x = \log_e x \quad (e = 2.71828\ldots)$$

II. Geometry

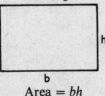

Rectangle:

Area $= bh$

Triangle:

Area $= \frac{1}{2}bh$

Circle:

Area $= \pi r^2$
Circumference $= 2\pi r$

3

Parallelogram:

Area = bh

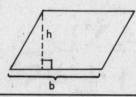

Trapezoid:

Area = $\frac{1}{2}h(b_1 + b_2)$

Distance between points (x_1, y_1) **and** (x_2, y_2) **in a plane is**

$$\sqrt{(x_1 - x_2)^2 + (y_1 - y_2)^2}$$

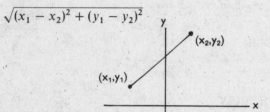

Distance between points (x_1, y_1, z_1) **and** (x_2, y_2, z_2) **in space is**

$$\sqrt{(x_1 - x_2)^2 + (y_1 - y_2)^2 + (z_1 - z_2)^2}$$

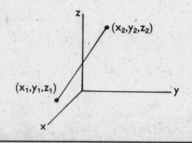

Pythagorean Theorem:

$x^2 + y^2 = z^2$

Triangle Inequality:

$|a - b| < c < a + b$

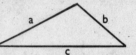

Rectangular solid:

Volume = $L \times W \times H$
Length of diagonal joining
opposite vertices = $\sqrt{L^2 + W^2 + H^2}$

Sphere:

Volume = $\frac{4}{3}\pi r^3$
Surface area = $4\pi r^2$

Prism, including Cylinder:

p = perimeter of base
A = area of base
h = altitude
Surface area (excluding ends) = ph
Volume = Ah

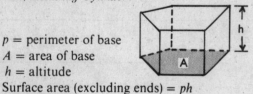

Pyramid, including Cone:

A = area of base
h = altitude
l = slant height
Volume = $\frac{1}{3}Ah$

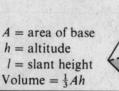

For the circular cylinder:

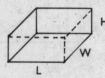

Surface area (excluding ends) = $2\pi rh$
Volume = $\pi r^2 h$

For the circular cone:

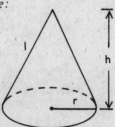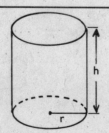

Surface area (excluding base) = πrl
Volume = $\frac{1}{3}\pi r^2 h$

Radian measure:

A *radian* is the measure of a central angle of a circle that intercepts an arc equal in length to the radius of the circle.

$2\pi^R = 360°$

$$1^R = \left(\frac{180}{\pi}\right)^°$$ (or approximately 57.3°)

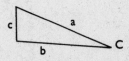

If θ is measured in radians, then

Area of sector $AOB = \frac{1}{2}r^2\theta$
Length of arc $AB = r\theta$.

III. Trigonometry

$\sin\theta = y/r$ $\csc\theta = 1/\sin\theta$ $\sin^2\theta + \cos^2\theta = 1$

$\cos\theta = x/r$ $\sec\theta = 1/\cos\theta$ $1 + \tan^2\theta = \sec^2\theta$

$\tan\theta = y/x$ $\cot\theta = 1/\tan\theta$ $1 + \cot^2\theta = \csc^2\theta$

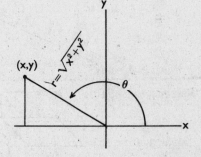

Degrees	0°	30°	45°	60°	90°	120°	135°	150°	180°	210°	225°	240°	270°	300°	315°	330°	360°
Radians	0	$\frac{\pi}{6}$	$\frac{\pi}{4}$	$\frac{\pi}{3}$	$\frac{\pi}{2}$	$\frac{2}{3}\pi$	$\frac{3}{4}\pi$	$\frac{5}{6}\pi$	π	$\frac{7}{6}\pi$	$\frac{5}{4}\pi$	$\frac{4}{3}\pi$	$\frac{3}{2}\pi$	$\frac{5}{3}\pi$	$\frac{7}{4}\pi$	$\frac{11}{6}\pi$	2π
$\sin\theta$	0	$\frac{1}{2}$	$\frac{\sqrt{2}}{2}$	$\frac{\sqrt{3}}{2}$	1	$\frac{\sqrt{3}}{2}$	$\frac{\sqrt{2}}{2}$	$\frac{1}{2}$	0	$-\frac{1}{2}$	$-\frac{\sqrt{2}}{2}$	$-\frac{\sqrt{3}}{2}$	-1	$-\frac{\sqrt{3}}{2}$	$-\frac{\sqrt{2}}{2}$	$-\frac{1}{2}$	0
$\cos\theta$	1	$\frac{\sqrt{3}}{2}$	$\frac{\sqrt{2}}{2}$	$\frac{1}{2}$	0	$-\frac{1}{2}$	$-\frac{\sqrt{2}}{2}$	$-\frac{\sqrt{3}}{2}$	-1	$-\frac{\sqrt{3}}{2}$	$-\frac{\sqrt{2}}{2}$	$-\frac{1}{2}$	0	$\frac{1}{2}$	$\frac{\sqrt{2}}{2}$	$\frac{\sqrt{3}}{2}$	1
$\tan\theta$	0	$\frac{\sqrt{3}}{3}$	1	$\sqrt{3}$	∞	$-\sqrt{3}$	-1	$-\frac{\sqrt{3}}{3}$	0	$\frac{\sqrt{3}}{3}$	1	$\sqrt{3}$	∞	$-\sqrt{3}$	-1	$-\frac{\sqrt{3}}{3}$	0

Law of Sines:

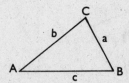

$$\frac{a}{\sin A} = \frac{b}{\sin B} = \frac{c}{\sin C}$$

Law of Cosines:

$$c^2 = a^2 + b^2 - 2ab\cos C$$

Addition and Subtraction Formulas:

$$\sin(A + B) = \sin A \cos B + \sin B \cos A$$

$$\cos(A + B) = \cos A \cos B - \sin A \sin B$$

$$\sin(A - B) = \sin A \cos B - \sin B \cos A$$

$$\cos(A - B) = \cos A \cos B + \sin A \sin B$$

$$\tan(A + B) = \frac{\tan A + \tan B}{1 - \tan A \tan B}$$

$$\tan(A - B) = \frac{\tan A - \tan B}{1 + \tan A \tan B}$$

Double Angle Formulas:

$$\sin 2A = 2 \sin A \cos A$$

$$\cos 2A = \cos^2 A - \sin^2 A$$

$$\tan 2A = \frac{2 \tan A}{1 - \tan^2 A}$$

$$= 2 \cos^2 A - 1$$

$$= 1 - 2 \sin^2 A$$

Half-Angle Formulas:

$$\sin \frac{A}{2} = \pm \sqrt{\frac{1 - \cos A}{2}}$$

$$\tan \frac{A}{2} = \pm \sqrt{\frac{1 - \cos A}{1 + \cos A}}$$

$$\cos \frac{A}{2} = \pm \sqrt{\frac{1 + \cos A}{2}}$$

$$= \frac{\sin A}{1 + \cos A}$$

$$= \frac{1 - \cos A}{\sin A}$$

Product Formulas:

$$\sin A + \sin B = 2 \sin\left(\frac{A + B}{2}\right) \cos\left(\frac{A - B}{2}\right)$$

$$\sin A - \sin B = 2 \cos\left(\frac{A + B}{2}\right) \sin\left(\frac{A - B}{2}\right)$$

$$\cos A + \cos B = 2 \cos\left(\frac{A + B}{2}\right) \cos\left(\frac{A - B}{2}\right)$$

$$\cos A - \cos B = -2 \sin\left(\frac{A + B}{2}\right) \sin\left(\frac{A - B}{2}\right)$$

IV. Conic Sections

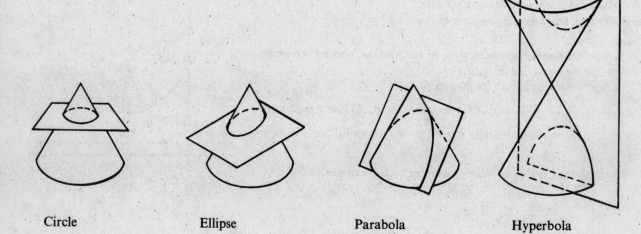

Circle   Ellipse   Parabola   Hyperbola

A circle is a set of points, in a plane, equidistant from a fixed point (center). If a circle has center (h, k) and radius r, its equation is

$$(x - h)^2 + (y - k)^2 = r^2.$$

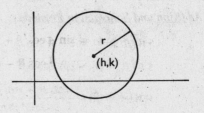

An *ellipse* is a set of points in a plane, the sum of whose distances from two fixed points (foci) is constant.
The most useful equation forms are:

$$\frac{(x - h)^2}{a^2} + \frac{(y - k)^2}{b^2} = 1 \quad \text{(major axis horizontal), and}$$

$$\frac{(x - h)^2}{b^2} + \frac{(y - k)^2}{a^2} = 1 \quad \text{(major axis vertical), where}$$

(h, k) = center,
$2a$ = length of major axis,
$2b$ = length of minor axis,
$c^2 = a^2 - b^2$, where $2c$ = distance between foci.

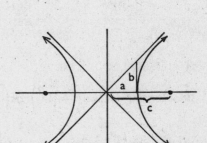

An *hyperbola* is the set of points in a plane such that the absolute value of the difference of the distance from two fixed points is a constant.
The most useful equation forms are

$$\frac{(x - h)^2}{a^2} - \frac{(y - k)^2}{b^2} = 1 \quad \left(\text{horizontal transverse axis, slope of asymptotes } \pm\frac{b}{a}\right)$$

$$\frac{(y - k)^2}{a^2} - \frac{(x - h)^2}{b^2} = 1 \quad \left(\text{vertical transverse axis, slope of asymptotes } \pm\frac{a}{b}\right)$$

(h, k) = center,
a is distance from center to vertex,
b is distance from vertex to asymptotes,
c is distance from center to focus, and $a^2 + b^2 = c^2$.

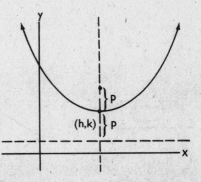

A *parabola* is a set of points, in a plane, equidistant from a fixed point (focus) and a fixed line (directrix).
The most useful equation forms are:

$$(x - h)^2 = 4p(y - k) \quad \text{(axis of symmetry is vertical line } x = h\text{), and}$$

$$(y - k)^2 = 4p(x - h) \quad \text{(axis of symmetry is horizontal line } y = k\text{), where}$$

(h, k) is the vertex;
p is the directed distance from the vertex to the focus.

Chapter 1

FUNCTIONS AND THEIR PROPERTIES

Section 1.1: Definitions Relating to Functions

Any set of ordered pairs is called a *relation*.
The set of all first coordinates of the ordered pairs is the *domain* of the relation.
The set of all second coordinates is the *range* of the relation.
A relation in which no two different ordered pairs have the same first coordinate is called a *function*.

Q1. The domain of the relation $\{(3, -5), (4, 12), (8, 8)\}$ is _____.

> Answer: $\{3, 4, 8\}$.

Q2. The range of the relation $\{(0, 4), (4, 5), (0, 3), (\frac{1}{2}, \sqrt{5})\}$ is _____.

> Answer: $\{\sqrt{5}, 3, 4, 5\}$.

Q3. If $f(x) = x^2 - x + 4$, then $f(-2) = \underline{(-2^2) - 2 + 4 = 4(2) + 4 = 10}$.

> Answer: $(-2)^2 - (-2) + 4 = 10$.

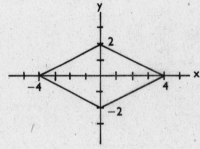

Q4. Given the relation shown on the graph.
  (a) The domain of the relation is __0, 4, 0, -4__.
  (b) The range of the relation is __2, 0, -2, 0__.
  (c) Is the relation a function? __no__.

> Answers: (a) $\{x: -4 \le x \le 4\}$.   (b) $\{y: -2 \le y \le 2\}$.   (c) No.

Q5. Given the relation shown on the graph.
   (a) The domain is ___-3 ≤ x ≤ 3___.
   (b) The range is ___0 ≤ y ≤ 9___.
   (c) Is the relation a function? ___yes___.

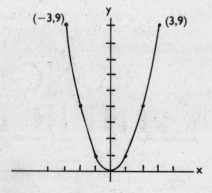

Answers: (a) $\{x: -3 \leq x \leq 3\}$.   (b) $\{y: 0 \leq y \leq 9\}$.   (c) Yes.

Authors' Note: It is often helpful to think of a numerical function as a rule that associates with a given number another unique number. For instance, if $f(x) = 2x + 1$, then the rule f says "take a number, multiply it by 2, then add 1." If the rule, f, is applied to the number 5, then it associates with 5 the unique number 11.

Note that if $g(x) = \pm\sqrt{x}$, then g is a rule that associates with each positive real number the two distinct square roots of the number. Hence g is not a function.

$\{x: a \leq x \leq b\}$ is often designated by $[a, b]$ ... a *closed* interval.
$\{x: a < x < b\}$ is often designated by (a, b) ... an *open* interval.

Problem Set 1.1

1. Which of the following graphs represent functions?
   (a)   (b)   (c)

 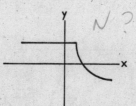 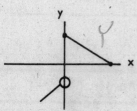

   (d)   (e)   (f)

 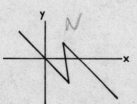 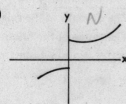

2. If $f(x) = 1 - x^2$, what is the range of f?

3. If $w(x) = 4x^2 - 8x + 3$, calculate $w(0) + w(\frac{1}{2}) + w(-1)$.

4. Graph $f(x) = \begin{cases} x & \text{if } x \geq 0 \\ 1 & \text{if } -1 < x < 0 \\ x - 2 & \text{if } x \leq -1 \end{cases}$
if the domain of f is $[-3, 3]$.
What is the range of f?

5. The symbol $[\![x]\!]$ is frequently used to designate the greatest integer less than or equal to x. Consider the function defined by $f(x) = [\![x]\!]$.
   (a) Calculate $f(5.43)$, $f(3)$, $f(\frac{1}{2})$, $f(\sqrt{2})$, and $f(-4.332)$.
   (b) What is the range of f, given that its domain is the set of real numbers?
   (c) Sketch a graph of f.

6. What real numbers are *not* in the domain of
$$f(x) = \frac{x-3}{x^2 + 3x + 2}?$$

7. What is the range for each function?
   (a) $f(x) = \sin x$
   (b) $g(x) = \tan x$
   (c) $h(x) = \log_{10} x$

Solutions and Comments for Problem Set 1.1

1. (b), (c), and (d) represent functions.

> *Vertical line test for functions:* If a vertical line never intersects a graph in more than one point, the graph represents a function. In (f), the vertical line $x = 0$ intersects the graph in two points.

> Note that the notation $[-5, -3] \cup [0, 3]$ is much more compact than $\{y: -5 \le y \le -3\} \cup \{y: 0 \le y \le 3\}$.

5. (a) $f(5.43) = 5$, $f(3) = 3$, $f(\frac{1}{2}) = 0$, $f(\sqrt{2}) = 1$, $f(-4.332) = -5$.
   (b) The set of all integers.
   (c)

2. $\{y: y \le 1\} = (-\infty, 1]$.

> Function represents a parabola with vertex $(0, 1)$, opening downward. Be able to read mathematical symbolism. For instance, the notation $\{y: y \le 1\}$ can be read "the set of all numbers less than or equal to 1." The variable y is a *dummy variable* since it does not have to be referenced when describing the set.

3. $3 + 0 + 15 = 18$.

> Function represents a parabola.

> *Common error:*
> $f(-4.332) = -4$.

4. Range of f is $[-5, -3] \cup [0, 3]$.

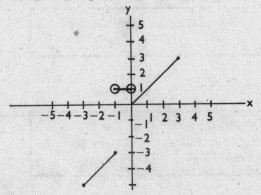

6. $f(x) = \dfrac{x-3}{(x+2)(x+1)}$. The domain of f does not include -2 or -1.

> *Common error:* f is undefined at $x = 3$.

7. (a) $[-1, 1]$;
   (b) $(-\infty, \infty)$;
   (c) $(-\infty, \infty)$.

> Remember that $\log_{10} x$ is an exponent. Any real number can be an exponent. Knowledge of the graphs of $y = \sin x$, $y = \tan x$, and $y = \log_{10} x$ is helpful in determining range values.

*Section 1.1A: Vector Functions and Parametrically Defined Functions

Points in a plane can be represented by expressing x and y in terms of a third variable, t. If f_1 and f_2 are functions such that

$$x = f_1(t) \quad \text{and} \quad y = f_2(t),$$

then these equations are called *parametric equations* and t is called a *parameter*.

If a, b, c, and d are real numbers, then the parametric equations

$$x = a + bt \quad \text{and} \quad y = c + dt$$

represent a line containing the point (a, c) with *direction vector* (b, d).

Q1. Describe the graph of the function defined by $x = 2t$, $y = 4t$. _____

> Answer: Eliminating t, one obtains $y = 2x$. The graph is a line through the origin with slope 2.

Authors' Note: The line in Q1 could be designated by the vector equation $(x, y) = (2t, 4t)$.

Q2. Describe the graph of the relation defined by $x = \cos t$, $y = \sin t$. _____

> Answer: Since $\cos^2 t + \sin^2 t = 1$, we have $x^2 + y^2 = 1$. The graph is a circle with center at the origin and radius 1.

Q3. Find parametric equations to represent the line containing $(1, 2)$ and $(4, 9)$. _____

> Answer: The direction vector is $(4 - 1, 9 - 2) = (3, 7)$. Hence $x = 1 + 3t$, $y = 2 + 7t$ is a parametric representation for the line. Another representation would be $x = 4 + 3t$, $y = 9 + 7t$.

Q4. What is the rectangular Cartesian equation for $x = t^3 - 1$, $y = t^6$? _____

> Answer: $t^3 = x + 1$; hence $y = (x + 1)^2$ is the desired equation.

Q5. In three-dimensional space, find parametric equations for the line containing $(1, 2, 3)$ and $(-2, 3, -8)$.
_____ $x = +3t$, $y = 2+t$, $z = 3-15$ $(-3, 1, -11)$

> Answer: The direction vector is $(-2 - 1, 3 - 2, -8 - 3) = (-3, 1, -11)$. Hence $x = 1 - 3t$, $y = 2 + t$, $z = 3 - 11t$ is a parametric representation for the line.

Authors' Note: Don't be careless when converting from parametric equations to rectangular coordinates. For instance, $x = \sin t$, $y = 3 \sin t$ can "carelessly" be converted to $y = 3x$, whereas, in fact, the function represented by the parametric equations is just a subset of the line $y = 3x$ since $-1 \le x \le 1$ and $-3 \le y \le 3$.

Problem Set 1.1A

1. Find the slope of the line represented by parametric equations $x = 3t + 1$, $y = t + 4$.

2. Find the rectangular Cartesian equations of the following:
   (a) $x = \frac{1}{2}t$, $y = 6t$
   (b) $x = t^3$, $y = t^6$
   (c) $x = t^4$, $y = t^8$
   (d) $x = t + 4$, $y = 3t - 6$
   (e) $x = 3 \sin t$, $y = 3 \cos t$
   (f) $x = 2 \sin t$, $y = 4 \cos t$
   (g) $x = \dfrac{1}{t}$, $y = t$
   (h) $x = 6t^2$, $y = 3$

3. Find parametric equations for
   (a) the line containing $(1, -4)$ with direction vector $(2, -7)$
   (b) the line containing $(-3, 2)$ and $(5, -11)$

4. In three-dimensional space, find parametric equations for
   (a) the line containing $(4, 1, -3)$ with direction vector $(1, 4, -8)$
   (b) the line containing $(1, 1, 1)$ and $(-3, -6, 4)$

5. Describe the graphs of the following vector functions:
   (a) $(x, y) = (4 + t, 6 + 2t)$
   (b) $(x, y) = (2t, t^2)$
   (c) $(x, y) = (t + 45, 7)$
   (d) $(x, y) = (t - 4, 4 - t)$

Solutions and Comments for Problem Set 1.1A

1. By letting $t = 0$ and $t = 1$, we can find that the line contains $(1, 4)$ and $(4, 5)$. Hence the slope is
$$\frac{5 - 4}{4 - 1} = \frac{1}{3}.$$

> The direction vector is $(3, 1)$ and the slope is merely the ratio of y-coordinate of direction vector to x-coordinate of direction vector.

2. (a) $y = 12x$ (line);
   (b) $y = x^2$ (parabola);
   (c) $y = x^2$, $x \ge 0$ (part of a parabola);
   (d) Solving for t, we obtain $x - 4 = \dfrac{y + 6}{3}$, which can be simplified to $y = 3x - 18$ (line);
   (e) $\dfrac{x^2}{9} + \dfrac{y^2}{9} = 1$, or $x^2 + y^2 = 9$ (circle);
   (f) $\dfrac{x^2}{4} + \dfrac{y^2}{16} = 1$ (ellipse);

(g) $y = \dfrac{1}{x}$ (hyperbola);

(h) $y = 3$, $x \geq 0$ (ray).

(c) *Common error:* $y = x^2$. Note that $x = t^4$ means $x \geq 0$. The parametric equations in (b) and (c) *do not* represent the same set of points.

(h) *Common error:* $y = 3$. Note that $x = 6t^2$ means $x \geq 0$.

3. (a) $x = 1 + 2t$, $y = -4 - 7t$;
   (b) $x = -3 + 8t$, $y = 2 - 13t$.

(b) The direction vector is
$(5 - (-3), -11 - 2) = (8, -13)$.

4. (a) $x = 4 + t$, $y = 1 + 4t$, $z = -3 - 8t$;
   (b) $x = 1 + 4t$, $y = 1 + 7t$, $z = 1 - 3t$.

(b) Direction vector is
$(1 - (-3), 1 - (-6), 1 - 4) = (4, 7, -3)$.

5. (a) A line ($y = 2x - 2$);
   (b) A parabola ($y = \frac{1}{4}x^2$);
   (c) A line ($y = 7$);
   (d) A line ($y = -x$).

Note in (c) that $x = t + 45$ means that x can assume any real number value. Contrast to 2(c) and 2(h).

Section 1.2: Functions: Sum, Difference, Product, Quotient, and Composition

If f and g are functions, then

$$(f + g)(x) = f(x) + g(x)$$

$$(f - g)(x) = f(x) - g(x)$$

$$(f \times g)(x) = f(x) \times g(x)$$

$$\left(\dfrac{f}{g}\right)(x) = \dfrac{f(x)}{g(x)} \quad \text{if } g(x) \neq 0$$

Q1. If $f(x) = x^2 - x - 2$ and $g(x) = x - 2$, then

(a) $(f + g)(x) = $ _____

(b) $(f - g)(x) = $ _____

(c) $(f \times g)(x) = $ _____

(d) $\left(\dfrac{f}{g}\right)(x) = $ _____

Answers: (a) $x^2 - 4$.   (b) $x^2 - 2x$.   (c) $x^3 - 3x^2 + 4$.

(d) $\dfrac{x^2 - x - 2}{x - 2} = \dfrac{(x - 2)(x + 1)}{x - 2} = x + 1$ if $x \neq 2$.

Q2. If $w(x) = \sqrt{x^2 + 4}$ and $t(x) = x^{1/3}$, then

(a) $(w + t)(8) = $ _____

(b) $(w \times t)(-1) = $ _____

Answers: (a) $w(8) + t(8) = \sqrt{68} + 8^{1/3} = 2\sqrt{17} + 2 = 2(1 + \sqrt{17})$.

(b) $w(-1) \times t(-1) = \sqrt{5}(-1) = -\sqrt{5}$.

The *composition* of f and g, denoted by $f \circ g$, is defined by $(f \circ g)(x) = f(g(x))$.

Q3. If $f(x) = x + 1$ and $g(x) = x^2 + 2$, then
    (a) $(f \circ g)(x) = $ _____
    (b) $(g \circ f)(x) = $ _____
    (c) $(f \circ f)(x) = $ _____
    (d) $(f \circ g \circ f)(x) = $ _____

> Answers: (a) $f(g(x)) = f(x^2 + 2) = (x^2 + 2) + 1 = x^2 + 3$.
> (b) $g(f(x)) = g(x + 1) = (x + 1)^2 + 2 = x^2 + 2x + 3$.
> (c) $f(f(x)) = f(x + 1) = (x + 1) + 1 = x + 2$.
> (d) $(f \circ g)(f(x)) = f(g(x + 1)) = f(x^2 + 2x + 3) = (x^2 + 2x + 3) + 1 = x^2 + 2x + 4$.

Authors' Notes: Don't confuse $(f \times g)(x)$ with $(f \circ g)(x)$.

A composition of functions is basically one rule followed by another rule. For instance, $f \circ g$ is a rule that says "take a number, square it, add 2, then (applying rule f) add 1."

Problem Set 1.2

1. If $f(x) = \dfrac{1}{x - 5}$ and $g(x) = x^3 + 2x$, calculate
    (a) $(f + g)(0)$
    (b) $(f \times g)(2)$
    (c) $(f - g)(5)$
    (d) $\left(\dfrac{f}{g}\right)(6)$

2. If $f(x) = 2x$, $g(x) = -x$ and $h(x) = 4$, find
    (a) $(f \circ g \circ h)(x)$
    (b) $(h \circ f \circ g)(x)$
    (c) $(g \circ h \circ f)(x)$

3. If $s(x) = \sqrt{4 - x}$ and $t(x) = x^2$, find
    (a) the domain of S
    (b) the domain of t
    (c) $s \circ t$
    (d) $t \circ s$

    (e) the domain of $s \circ t$
    (f) the domain of $t \circ s$

4. If $f(x) = m_1 x + b_1$ and $g(x) = m_2 x + b_2$, find
    (a) $(f \circ g)(x)$
    (b) $(f \circ g)(0)$

5. If $F(x) = \dfrac{3}{x^2 - 4}$ and $I(x) = x$, find
    (a) $(F \times I)(x)$
    (b) $(F \circ I)(x)$
    (c) $(I \circ F)(x)$

6. Given that $s(x) = \cos x$, where $x \in [0°, 360°]$ and $L(x) = \log_{10} x$. What is the domain and range for the composite function $(L \circ s)(x)$?

7. If $P(x) = x^2 + 3x$, find $P(x + 2)$.

Solutions and Comments for Problem Set 1.2

1. (a) $f(0) + g(0) = -\frac{1}{5} + 0 = -\frac{1}{5}$;
    (b) $f(x) \times g(2) = (-\frac{1}{3})(12) = -4$;
    (c) $(f - g)(5) = f(5) - g(5)$ is not defined since $f(5)$ is not defined,
    (d) $\dfrac{f(6)}{g(6)} = \dfrac{1}{228}$.

> If two functions are combined with one of the operations $+$, $-$, $\times$, or $\div$, the domain of the resulting function is the intersection of the two domains. In part (c), 5 is not in the domains of f and g.

2. (a) $f(g(h(x))) = f(g(4)) = f(-4) = -8;$
   (b) $h(f(g(x))) = h(f(-x)) = h(-2x) = 4;$
   (c) $g(h(f(x))) = g(h(2x)) = g(4) = -4.$

> The rule h merely says "assign to any number the number 4." Hence the rule is easy to apply. Since h is the last rule applied in (b), the result must be 4 since the domain for all of the functions and their various compositions is the set of real numbers.

3. (a) $D_s = \{x : x \le 4\};$
   (b) $D_t = $ the set of real numbers;
   (c) $(s \circ t)(x) = s(x^2) = \sqrt{4 - x^2};$
   (d) $(t \circ s)(x) = t(\sqrt{4 - x}) = (\sqrt{4 - x})^2 = 4 - x;$
   (e) $D_{s \circ t} = \{x : -2 \le x \le 2\};$
   (f) $D_{t \circ s} = \{x : x \le 4\}.$

> In general, if f and g are functions, then the domain of $f \circ g$ must be a subset of the domain of g (since g is the first rule that must be applied). If a number c is not in the domain of g, then $(f \circ g)(c)$ is not defined.
>
> *Common error:* $(\sqrt{4 - x})^2 = 4 - x.$
>
> Note that in (d), $(\sqrt{4 - x})^2$ is equal to $4 - x$ since $D_{t \circ s}$ does not contain any numbers greater than 4. However, if the domain consisted of all real numbers, it would be important to realize that $(\sqrt{4 - x})^2 = |4 - x|$. Note that $y = 4 - x$ and $y = |4 - x|$ represent different functions over the set of real numbers.

4. (a) $f(m_2 x + b_2) = m_1(m_2 x + b_2) + b_1$
   $\qquad = m_1 m_2 x + m_1 b_2 + b_1;$
   (b) $(f \circ g)(0) = m_1 b_2 + b_1.$

> The composition of two linear functions always produces a linear function.

5. (a) $(F \times I)(x) = \dfrac{3x}{x^2 - 4};$
   (b) $F(I(x)) = F(x) = \dfrac{3}{x^2 - 4};$
   (c) $I(F(x)) = I\left(\dfrac{3}{x^2 - 4}\right) = \dfrac{3}{x^2 - 4}.$

> The function $I(x) = x$ is the *identity* element in composition of functions. That is, $f \circ I = I \circ f = f$ for any function f. I is a rule that says "assign each number to itself."

6. $D_{L \circ s} = [0°, 90°) \cup (270°, 360°]$ and $R_{L \circ s} = (-\infty, 0].$

> $L(s(x)) = \log_{10}(\cos x)$. This requires $\cos x > 0$; hence $x \in [0°, 90°) \cup (270°, 360°]$, where $0 < \cos x \le 1$. Remember that $\log_{10} 1 = 0$ and, if $0 < c < 1$, then $\log_{10} c < 0$. Hence the range for the composite function is $(-\infty, 0]$.

7. $(x + 2)^2 + 3(x + 2) = x^2 + 7x + 10.$

> *Common error:* $P(x + 2) = P(x) + P(2).$

Section 1.3: Absolute Value and Inequalities

Definitions of *absolute value*:

(1) $|x| = \begin{cases} x & \text{if } x \ge 0 \\ -x & \text{if } x < 0 \end{cases}$
(2) $|x| = \max\{x, -x\}$
(3) $|x| = \sqrt{x^2}.$

Authors' Notes: If a point on a number line has coordinate x, then $|x|$ is the distance of the point from the origin.

If points A and B have coordinates x and y, respectively, then $|x - y|$ is the length of the segment AB.

Properties of absolute value:

(1) $|x - y| = |y - x|$ (2) $|x + y| \leq |x| + |y|$ (The Triangle Inequality);

(3) $|x||y| = |xy|$ (4) $|x| \geq 0$.

Authors' Notes: $|x| = -x$ is perfectly meaningful if $x < 0$.

If $b > 0$, then $|x - a| < b$ means $a - b < x < a + b$. Geometrically, we are describing the set of all points whose distance from a point with coordinate a is less than b.

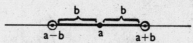

If $b > 0$, then $|x - a| > b$ means $x > a + b$ or $x < a - b$. Geometrically, the set of all points whose distance from a point with coordinate a is greater than b.

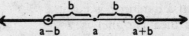

Don't confuse "and" and "or." Remember, "and" means intersection ($\cap$) and "or" means union ($\cup$).

Q1. Solve (a) $|x| < 5$ _____ (b) $|x| = 3$ _____

(c) $|x + 2| > 2$ _____ (d) $|x - 2| < -3$ _____

(e) $|x - 2| > -3$ _____ (f) $-2|x| > -4$ _____

Answers: (a) $-5 < x < 5$. (b) $x = -3$ or 3. (c) $x > 0$ or $x < -4$.

(d) empty set. (e) all reals. (f) $|x| < 2 \Rightarrow -2 < x < 2$.

Q2. Write "x is within b units of a" using absolute value notation. _____

Answer: $|x - a| < b$.

Q3. Graph the solution set for $|x| = |y - 2|$ on the grid provided (page 18). _____

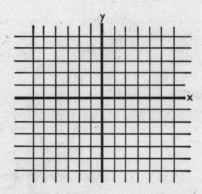

Answer: $x = y - 2$ or $x = -y + 2$. The graph is two intersecting
lines.

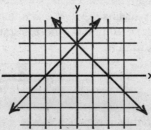

An *interval* is a subset of real numbers between two endpoints.
  A *closed interval* contains the endpoints, an *open interval* does not.
  A *neighborhood* of a real number c is an open interval containing c.
  A *deleted neighborhood* of c is a neighborhood of c *not* containing c.
  If c is the midpoint of an open interval (a, b), the interval is called a *neighborhood* of c with *radius* $|b - c|$.

Q4. Express with absolute value symbolism.
  (a) x is in a neighborhood of a with radius b. _____
  (b) x is in a deleted neighborhood of p with radius q. _____

Answers: (a) $|x - a| < b$.   (b) $0 < |x - p| < q$.

Problem Set 1.3

1. Solve $|3x + 2| \leq 5$.

2. Solve $|2x - 1| > 3$.

3. Solve $|2x - 1| < 3x - 3$.

4. Solve $|x| + |x - 2| = 4$.

5. Graph $|x| + |y| = 2$.

6. Solve $\dfrac{1}{x} < 1$.

Solutions and Comments for Problem Set 1.3

1. $-5 \leq 3x + 2 \leq 5 \Rightarrow -7 \leq 3x \leq 3$
    $\Rightarrow -7/3 \leq x \leq 1$, or $[-7/3, 1]$.

The closed interval describes a geometric
segment of length $3\frac{1}{3}$.

2. $2x - 1 > 3$ or $2x - 1 < -3$
$\Rightarrow x > 2$ or $x < -1$.

> A careless and common mistake is to write $2 < x < -1$. This means "2 is less than x *and* x is less than -1." This compound sentence describes the empty set.

3. $2x - 1 < 3x - 3$ and $2x - 1 > -(3x - 3)$
$\Rightarrow x > 2$ and $x > 4/5$. Hence $x > 2$.

> Contrast to problem #1. The variable on the right hand side of the original inequality makes it necessary to examine two separate cases.

4. If $x < 0$, then $-x - (x - 2) = 4 \Rightarrow x = -1$;
If $0 \le x < 2$, then $x - (x - 2) = 4$ (no solution);
If $x \ge 2$, then $x + x - 2 = 4 \Rightarrow x = 3$. Hence, $x = -1$ or $x = 3$.

> Three cases must be considered. A common mistake is to just "forget" about the absolute value bars. This error produces just one of the solutions. Absolute value has meaning . . . you can't just ignore it.

5. $x + y = 2$ in Quadrant I;
$-x + y = 2$ in Quadrant II;
$-x - y = 2$ in Quadrant III;
$x - y = 2$ in Quadrant IV.

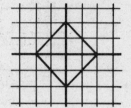

> By inspection, you can note that the solution must be contained by the rectangular region described by $-2 \le x \le 2$ and $-2 \le y \le 2$.

6. If $x > 0$, then $\dfrac{1}{x} < 1 \Rightarrow x > 1$. If $x < 0$, then $\dfrac{1}{x} < 1 \Rightarrow x < 1$. Hence, solution is $x > 1$ or $x < 0$.

> A very common mistake is to forget that x, a variable, can be negative. Note that if $x < 0$, then the left hand side of the inequality is negative and the inequality is true.

Section 1.4: Inverses of Functions

> If f is a one-to-one function, then the *inverse* of f is a function denoted by f^{-1} with the following properties:
> (i)   The domain of f^{-1} is the range of f.
> (ii) The range of f^{-1} is the domain of f.
> (iii) If $f(a) = b$, then $f^{-1}(b) = a$.

Authors' Notes: Using (iii), $(f^{-1} \circ f)(x) = (f \circ f^{-1})(x) = x$; that is, $f^{-1} \circ f = f \circ f^{-1} = I$, where I is the identity element for the composition of functions.
    If f has an inverse, and if $(a, b) \in f$, then $(b, a) \in f^{-1}$. That is, the inverse of f is the set of ordered pairs obtained by interchanging the coordinates of the ordered pairs of f. Hence the graphs of $y = f(x)$ and $y = f^{-1}(x)$ are symmetric with respect to the line $y = x$.

Q1. Which of the following functions (page 20) have inverses? _____

(a)   (b)   (c)   (d)

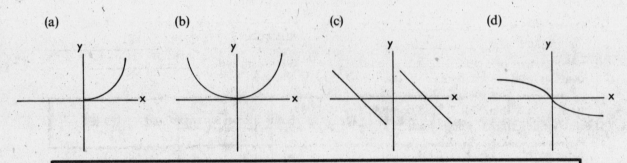

> **Answer:** (a) and (d). The graphs in (b) and (c) do not represent one-to-one functions; that is, there are values in the range which correspond to two different values in the domain.

Authors' Note: If a horizontal line intersects the graph of a function in more than one point, then the function does not have an inverse.

Q2. If $y = f(x) = 3x - 2$, then $f^{-1}(x) =$ _____ and $f^{-1}(8) =$ _____ .

> **Answer:** Interchange x and y, obtaining $x = 3y - 2$. Solving for y, one obtains $y = \dfrac{x + 2}{3}$.
>
> Hence $f^{-1}(x) = \dfrac{x + 2}{3}$ and $f^{-1}(8) = \frac{10}{3}$. (Note that $f(\frac{10}{3}) = 8$.)

Authors' Note: In Q2, f is a rule that says "take a number, multiply it by 3, then subtract 2." Hence, to "undo" the rule f, the rule f^{-1} must say "take a number, add 2, then divide by 3."

With restrictions as noted, the trigonometric functions sin, cos, and tan have inverse functions defined as follows:

$$\sin^{-1} x = \arcsin x = \{(x, y): \sin y = x \text{ and } -\tfrac{\pi}{2} \le y \le \tfrac{\pi}{2}\}.$$

$$\cos^{-1} x = \arccos x = \{(x, y): \cos y = x \text{ and } 0 \le y \le \pi\}.$$

$$\tan^{-1} x = \arctan x = \{(x, y): \tan y = x \text{ and } -\tfrac{\pi}{2} < y < \tfrac{\pi}{2}\}.$$

Authors' Note: It is *extremely important* to realize that the familiar function sin has no inverse and that $\sin^{-1}$ or arc sin denotes the inverse of a particular restriction of sin. *Be very aware* that $\sin(\sin^{-1} x) = x$ for any x in the domain of $\sin^{-1}$, but $\sin^{-1}(\sin y) = y$ if and only if $y \in [-\tfrac{\pi}{2}, \tfrac{\pi}{2}]$.

Q3. (a) $\arc\sin\frac{1}{2} =$ _____

(c) $\arctan 1 =$ _____

(e) $\arctan(-\sqrt{3}) =$ _____

(b) $\arccos\left(-\frac{\sqrt{2}}{2}\right) =$ _____

(d) $\arc\sin 2 =$ _____

Answers: (a) $\frac{\pi}{6}$. (b) $\frac{3\pi}{4}$. (c) $\frac{\pi}{4}$. (d) $\arc\sin 2$ is undefined. (e) $-\frac{\pi}{3}$.

 Common error: $\arctan(-\sqrt{3}) = \frac{2\pi}{3}$.

Authors' Note: One can read $\arctan(-\sqrt{3})$ as "an angle whose tangent is $-\sqrt{3}$." However, while there are an infinity of angles whose tangents are $-\sqrt{3}$, there is only one $\arctan(-\sqrt{3})$. Remember that $-\frac{\pi}{2} < \arctan x < \frac{\pi}{2}$.

 Common error: $f^{-1} = \frac{1}{f}$.

Authors' Note: If f is a function, then f^{-1} refers to the inverse function (if it exists) and *not* a multiplicative inverse. Watch your use and reading of notation. If $f(x) = x + 2$, then $f^{-1}(x) = x - 2$ and

$$[f(x)]^{-1} = \frac{1}{x + 2}.$$

Problem Set 1.4

1. If $g = \{(0, 1), (1, 2), (2, 3), (3, 4)\}$, what is g^{-1}?

2. Consider $f(x) = x^2 + 2$. Does f have an inverse? Justify your response.

3. Find f^{-1} for each of the following functions.
   (a) $f(x) = x + 6$
   (b) $f(x) = \frac{x}{7}$
   (c) $f(x) = x^2 + 5, x \geq 0$
   (d) $f(x) = \sqrt{x + 2}, x \geq -2$
   (e) $f(x) = \frac{6}{x + 1}$
   (f) $f(x) = x^3$

4. If $f(x) = 2x + 7$, find
   (a) $f^{-1}(x)$
   (b) $f^{-1}(0)$
   (c) $(f^{-1} \circ f)(x)$
   (d) $(f \circ f^{-1})(3)$

5. Find
   (a) $\arc\sin 1$
   (b) $\arccos 0$
   (c) $\arctan\frac{\sqrt{3}}{3}$
   (d) $\arc\sin\left(-\frac{\sqrt{2}}{2}\right)$
   (e) $\arccos\left(-\frac{1}{2}\right)$
   (f) $\arctan(-1)$
   (g) $\arccos\left(-\frac{\sqrt{3}}{2}\right)$
   (h) $\sin\left(\arccos\left(-\frac{\sqrt{3}}{2}\right)\right)$
   (i) $\tan(\arc\sin(-\frac{1}{2}))$

6. Sketch the graph of
   (a) $y = \arc\sin x$
   (b) $y = \arccos x$

Solutions and Comments for Problem Set 1.4

1. $g^{-1} = \{(1,0),(2,1),(3,2),(4,3)\}$.

> g^{-1} is obtained by interchanging the x- and y-coordinates in g.

2. f has no inverse since it is not a one-to-one function. For example, $f(1) = 3$ and $f(-1) = 3$.

> Interchanging x and y, we obtain $x = y^2 + 2$, which implies $y = \pm\sqrt{x-2}$. This relation is not a function.

3. (a) $f^{-1}(x) = x - 6$;
   (b) $f^{-1}(x) = 7x$;
   (c) $f^{-1}(x) = \sqrt{x-5}$, $x \geq 5$; (see graph)
   (d) $f^{-1}(x) = x^2 - 2$, $x \geq 0$;
   (e) $f^{-1}(x) = \dfrac{6-x}{x}$;
   (f) $f^{-1}(x) = x^{1/3}$ (or $\sqrt[3]{x}$).

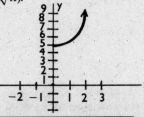

graph
for (c)

> Note that in (c), without the condition $x \geq 0$, f would have no inverse. $D_f = [0,\infty]$ and $R_f = [5,\infty]$. $x = y^2 + 5 \Rightarrow y = \sqrt{x-5}$. Note that $y = -\sqrt{x-5}$ is not considered since $D_f - 1 = R_f = [5,\infty]$. In (d), $x = \sqrt{y+2} \Rightarrow f^{-1}(x) = y = x^2 - 2$, where $x \geq 0$.
>
> *Common error:* $D_f - 1 =$ the set of real numbers. Note that $D_f - 1 = R_f = [0,\infty]$.

4. (a) $f^{-1}(x) = \dfrac{x-7}{2}$;
   (b) $-7/2$;
   (c) x;
   (d) 3.

> Note that $f(-\frac{7}{2}) = 0$, as expected. The answers for (c) and (d) require no computation since $f^{-1}\circ f = f\circ f^{-1} = I$, the identity function for composition.

5. (a) $\pi/2$;
   (b) $\pi/2$;
   (c) $\pi/6$;
   (d) $-\pi/4$;
   (e) $2\pi/3$;
   (f) $-\pi/4$;
   (g) $5\pi/6$;
   (h) $1/2$;
   (i) $-\sqrt{3}/3$.

> Remember that $\arcsin x$, $\arccos x$, and $\arctan x$ have restricted values.
>
> *Common errors:* $\arccos 0 = 0$. $\arcsin\left(-\dfrac{\sqrt{2}}{2}\right) = \dfrac{7\pi}{4}$.

6. (a) The solid graph is $y = \arcsin x$. The dotted graph is $y = \sin x$ on the interval $[-\frac{\pi}{2}, \frac{\pi}{2}]$. The dashed line is $y = x$, the line of symmetry for a function and its inverse.

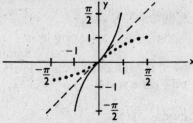

(b) Similar to (a), except that $y = \cos x$ is graphed on the interval $[0, \pi]$.

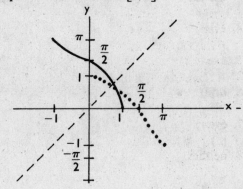

Section 1.5: Odd and Even Functions

A function is *odd* if $f(-x) = -f(x)$ for all x.
A function is *even* if $f(-x) = f(x)$ for all x.

Q1. Identity the function as odd, even, or neither.

(a) $f(x) = x^3$ _____

(b) $f(x) = x^3 + 1$ _____

(c) $f(x) = \sin x$ _____

(d) $f(x) = \cos x$ _____

(e) $f(x) = x^2 + 2x + 1$ _____

(f) $f(x) = x^4 - 2x^2$ _____

(g) $g(x) = x^5 - x^3 - x$ _____

> Answers: Odd functions (a, c, g).
> Even functions (d, f).
> Neither (b, e).

Authors' Note: An odd function is symmetric with respect to the origin.
An even function is symmetric with respect to the y-axis.

Problem Set 1.5

In problems 1–8, identify the function as odd, even, or neither.

1. $s(x) = \sin 2x$

2. $f(x) = x^2 + 1$

3. $f(x) = x^5 + 1$

4. $c(x) = \cos^2 x$

5. $g(x) = x^4 - x^2 + 2$

6. $f(x) = x(x^2 - 1)$

7. $f(x) = \dfrac{1 + |x|}{x^2}$

8. $f(x) = 5^x$

9. What is the result if you take the product of
   (a) two even functions?
   (b) two odd functions?

Solutions and Comments for Problem Set 1.5

1. odd, since $\sin(-2x) = -\sin 2x$;

2. even;

3. neither;

4. even;

5. even;

6. odd;

7. even;

8. neither.

In a polynomial function, if the exponents of the variable are all odd, then the function is odd since $(-x)^n = -x^n$ when n is odd. If the exponents of the variable are all even, then the function is even since $(-x)^n = x^n$ if n is even.

 Common error: $f(x) = x^5 + 1$ is odd.

Note that $x^5 + 1 = x^5 + 1 \cdot x^0$, and that 0 is an even integer. Also ... note that problems 1, 4, 7, and 8 are not polynomial functions.

9. In both cases, the product is an even function.

In (b), if f and g are odd, then let $h(x) = f(x)g(x)$. Then $h(-x) = f(-x)g(-x) = [-f(x)][-g(x)] = f(x)g(x) = h(x)$.

Section 1.6: Periodicity of Functions

If f is a function and if there exists a number $p > 0$ such that

$$f(x + p) = f(x)$$

for all x in the domain of f, then f is said to be *periodic*. If p is the smallest number with this property, then p is the *period* of f.

Q1. Which of the functions shown are periodic? _____

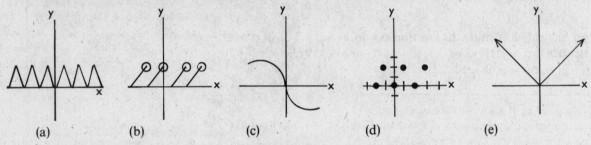

(a)   (b)   (c)   (d)   (e)

Answer: (a), (b), and (d).

Q2. What is the period of $f(x) = \sin x$? _____

Answer: Since $\sin(x + 2\pi) = \sin x$, the period is 2π.

Authors' Note: $\sin(x + 4\pi) = \sin x$, but 2π is the *smallest* positive number with the desired property.

Q3. What is the period of $g(x) = \tan x$? _____

Answer: π, since $\tan(x + \pi) = \tan x$.

Problem Set 1.6

1. If f and g are periodic functions with period 4 such that $f(2) = 6$ and $g(6) = 10$, calculate
   (a) $f(6)$
   (b) $g(-2)$
   (c) $(f + g)(14)$
   (d) $(f \times g)(-6)$
   (e) $(f \circ g)(18)$
   (f) $(g \circ f)(26)$

2. If the domain of f is the set of integers and if $f(x) = \begin{cases} 3 & \text{if } x \text{ is even} \\ 2 & \text{if } x \text{ is odd} \end{cases}$, is the function f periodic? If so, what is its period?

3. Identify each function as periodic or non-periodic. If periodic, state the period.
   (a) $g(x) = \cos x$
   (b) $i(x) = [\![x]\!]$ (greatest integer function)
   (c) $t(x) = \begin{cases} 1 & \text{if } x \geq 0 \\ -1 & \text{if } x < 0 \end{cases}$
   (d) $s(x) = 5 + \sin x$
   (e) $h(x) = 2^x$
   (f) $z(x) = \cos x + \sin x$
   (g) $c(x) = |\sin x|$

4. Identify the period for each function.
   (a) $f(x) = \sec x$
   (b) $g(x) = \cot x$
   (c) $h(x) = \csc x$

Solutions and Comments for Problem Set 1.6

1. (a) 6;
   (b) 10;
   (c) $f(14) + g(14) = 6 + 10 = 16$;
   (d) $f(-6)g(-6) = 6 \times 10 = 60$;
   (e) $f(g(18)) = f(10) = 6$;
   (f) $g(f(26)) = g(6) = 10$.

> The sum of two periodic functions with the same period defined over the real numbers is a periodic function with the same period. The same can be said for the product of such functions.

2. f is periodic; period $= 2$.

> If x is even, then $f(x) = f(x + 2) = 3$.
> If x is odd, then $f(x) = f(x + 2) = 2$.

3. (a) periodic, 2π;
   (b) non-periodic;
   (c) non-periodic;
   (d) periodic, 2π;
   (e) non-periodic;
   (f) periodic, 2π;
   (g) periodic, π.

> The comment for problem #1 relates to (f). For (g), note the effect of the absolute value bars.

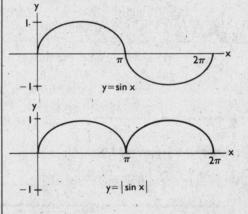

4. (a) 2π;
   (b) π;
   (c) 2π.

> These functions have the same period as their reciprocal functions. Note, however, that the domains are not the same as those of the respective reciprocal function. For instance, the domain of $y = \sec x$ does not include those values for which the value of $\cos x$ is zero.

Section 1.7: Symmetry and Asymptotes

Given a function f,

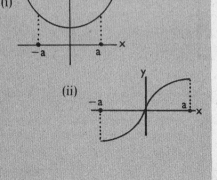
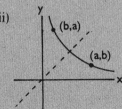

   (i)   If $f(-x) = f(x)$, then f is symmetric with respect to the y-axis.

   (ii) If $f(-x) = -f(x)$, then f is symmetric with respect to the origin.

  (iii) If for all $(a, b) \in f$, we have $(b, a) \in f$, then f is symmetric with respect to the line $y = x$.

Authors' Notes: When discussing symmetry, phrases like "with respect to the y-axis," "about the y-axis," and "around the y-axis" all mean the same thing.

An odd function is symmetric about the origin.

An even function is symmetric around the y-axis.

Note that symmetry with respect to the x-axis could certainly exist, but a relation that has such symmetry would not be a function.

A function that is symmetric about $y = x$ is its own inverse. That is, $f^{-1} = f$.

There are many useful relations, such as $x^2 + y^2 = 9$, which have lots of symmetry. This relation, a circle, is symmetric with respect to the x-axis, the y-axis, the origin, and the line $y = x$.

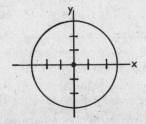

Q1. Identify the symmetry for each function.

  (a) $f(x) = x^2 + 1$ _____   (b) $f(x) = x^3$ _____

  (c) $f(x) = \dfrac{1}{x^2 - 1}$ _____   (d) $f(x) = \dfrac{1}{x}$ _____

Answers: (a) $f(-x) = f(x) \Rightarrow$ symmetry about the y-axis.

         (b) $f(-x) = -f(x) \Rightarrow$ symmetry with respect to the origin.

         (c) $f(-x) = f(x) \Rightarrow$ symmetry around the y-axis.

         (d) Also, f is symmetric about the origin. $(a, b) \in f \Rightarrow (b, a) \in f$. Hence, f is symmetric about $y = x$.

An *asymptote* is a line toward which a graph tends as x approaches some specific value, or as x increases or decreases without bound.

  If $f(x) \to \infty$ or $f(x) \to -\infty$ as $x \to a$, then the line $x = a$ is a *vertical asymptote*.

  If $f(x)$ approaches the constant value c as x increases or decreases without bound, then $y = c$ is a *horizontal asymptote*.

If the graph of f approaches a non-horizontal and non-vertical line as $x \to \infty$ or $x \to -\infty$, then the line is a *slant asymptote* for the function.

Authors' Notes: Values for which the denominator of a function is zero often determine vertical asymptotes.

Q2. Sketch a graph for each of the following functions, indicating all asymptotes.

(a) $f(x) = \dfrac{2}{x-2}$

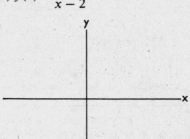

(b) $f(x) = \dfrac{x^2 + 1}{x}$

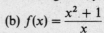

Answers: (a) Vertical asymptote: $x = 2$.
Horizontal asymptote: $y = 0$.

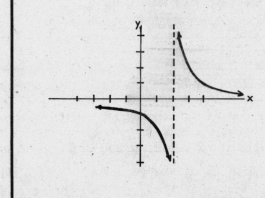

(b) $f(x) = \dfrac{x^2 + 1}{x} = x + \dfrac{1}{x}$

Vertical asymptote: $x = 0$.
Slant asymptote: $y = x$.

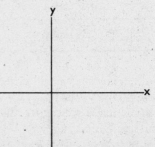

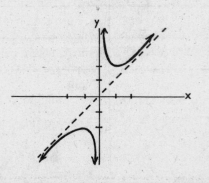

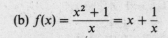

Authors' Notes: Writing functions in different forms is often helpful. For instance, in (b), writing $f(x) = x + \dfrac{1}{x}$, we can see that $f(x)$ gets "very close" to x as $x \to \infty$ or $x \to -\infty$.

Problem Set 1.7

In problems 1–10, identify symmetry and asymptotes.

1. $y = \dfrac{1}{x-1}$

2. $y = \dfrac{x}{x^2 - 4}$

3. $y = \dfrac{x^2}{x^2 - 1}$

4. $y = x^3 + x^2 - 3x - 1$

5. $f(x) = |x| + 1$

6. $f(x) = x^5 + x^3 + x$

7. $x^2 - y^2 = 1$

8. $y = \dfrac{3x^2}{2x^2 - 3x + 3}$

9. $f(x) = x^3 + 1$

10. $f(x) = \dfrac{5x^2 - 5x + 1}{x - 1}$

11. Find the asymptotes for the following hyperbolas:

   (a) $\dfrac{x^2}{9} - \dfrac{y^2}{16} = 1$

   (b) $xy = 4$

Solutions and Comments for Problem Set 1.7

	Asymptotes	Symmetry	Comments
1.	$x = 1$, $y = 0$	$y = x - 1$	This is the hyperbola $xy = 1$ shifted one unit to the right.
2.	$x = 2$, $x = -2$, $y = 0$	origin	In a quotient of polynomials, if degree of numerator is less than degree of denominator, then limit as $x \to \infty$ is 0.
3.	$x = 1$, $x = -1$, $y = 1$	y-axis	An even function.
4.	None	None	
5.	None	y-axis	An even function.
6.	None	origin	An odd function.
7.	$y = x$, $y = -x$	x-axis, y-axis, origin	See comment for problem #11.
8.	$y = \frac{3}{2}$	None	Discriminant of denominator is negative; hence denominator cannot be zero for any value of x. Note that if numerator and denominator are divided by x^2, we get $y = \dfrac{3}{2 - \dfrac{3}{x} + \dfrac{3}{x^2}}$. This approaches $3/2$ as $x \to \infty$ or $x \to -\infty$.
9.	None	See comment.	Graph is $y = x^3$ shifted up one unit. It would be symmetric to the point $(0, 1)$.
10.	$x = 1$, $y = 5x$	See comment.	Write as $y = 5x + \dfrac{1}{x - 1}$. One could establish symmetry with respect to the point $(1, 5)$.

11. (a) $y = \frac{4}{3}x$ and $y = -\frac{4}{3}x$;   (b) $y = 0$ and $x = 0$.

> Asymptotes for hyperbola $\dfrac{x^2}{a^2} - \dfrac{y^2}{b^2} = 1$ are $y = \dfrac{b}{a}x$ and $y = -\dfrac{b}{a}x$.

Section 1.8: Zeros of a Function

If $y = f(x)$ and if c is a number such that $f(c) = 0$, then c is a *root* or *zero* of the equation $f(x) = 0$. Roots can be real or imaginary if the domain is the set of complex numbers. Any real roots are x-intercepts on the graph of $y = f(x)$. Consider now a polynomial function

$$P(x) = a_n x^n + a_{n-1} x^{n-1} + \cdots + a_1 x + a_0$$

where the coefficients $a_n, a_{n-1}, \ldots, a_1, a_0$ are real numbers and $a_n \neq 0$. There are several important theorems relating to such a function.

(1) *Factor Theorem:* r is a root of $P(x) = 0$ if and only if $x - r$ is a factor of $P(x)$.

(2) *Remainder Theorem:* The remainder in the division of $P(x)$ by $x - r$ is equal to $P(r)$.

(3) *Rational Root Theorem:* If $P(x)$ has integer coefficients and if the rational number p/q, in lowest terms, is a root of $P(x) = 0$, then p is an integer factor of a_0 and q is an integer factor of a_n.

(4) If n is an odd integer, then there is at least one real root for the equation $P(x) = 0$.

(5) Complex roots of $P(x) = 0$ occur in conjugate pairs. That is, if $a + bi$ is a complex number root, then $a - bi$ must also be a root.

(6) If the terms of the polynomial $P(x)$ are arranged in descending order of powers, then

    (i) the number of positive real roots of $P(x) = 0$ is equal to the number of sign changes in $P(x)$ or is less than this by a positive even integer number.

    (ii) the number of negative real roots is equal to the number of sign changes in $P(-x)$ or is less than this by a positive even integer number.

Authors' Notes: 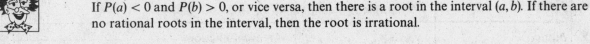 Remember the premise for the theorems above states that the coefficients in the polynomial are real numbers.

If $P(a) < 0$ and $P(b) > 0$, or vice versa, then there is a root in the interval (a, b). If there are no rational roots in the interval, then the root is irrational.

Q1. Find all complex roots for the polynomial functions.

    (a) $x(3x + 2)(2x - 1)(x^2 + 2) = 0$ _____

    (b) $2x^2 - 3x + 5 = 0$ _____

> Answers: (a) Setting each factor equal to zero, we find that the roots are $0, -2/3, 1/2, \sqrt{2}i$, and $-\sqrt{2}i$.
>
>   (b) Using the quadratic formula, the two roots are $\dfrac{3 + \sqrt{31}i}{4}$ and $\dfrac{3 - \sqrt{31}i}{4}$.

Q2. List all of the possible rational roots of $3x^2 + 4x^4 - 3x^3 + 2x - 6 = 0$. _____

> Answer: If p/q is a rational root, then p must be a factor of 6 and q must be a factor of 3. Hence p has possible values $\pm 1, \pm 2, \pm 3$, and ± 6, while q has possible values $\pm 1, \pm 3$. Hence the possible rational roots are $\pm 1, \pm 2, \pm 3, \pm 6, \pm 1/3, \pm 2/3$.

Q3. Discuss the nature of the roots of $x^7 - 3x^4 - 5x^3 + 2x^2 - 3x + 1 = 0$. _____

> Answer: There is at least one real root and at most seven real roots. With four sign changes in $P(x)$, there are 4, 2, or 0 positive real roots. $P(-x) = -x^7 - 3x^4 + 5x^3 + 2x^2 + 3x + 1$. Since there is one sign change, there is 1 negative real root. The only possible rational roots are 1 and -1. There are either 2, 4, or 6 imaginary roots.

Q4. Are either of $x - 1$ or $x + 1$ factors of $P(x) = 3x^4 - 2x^2 + 2x + 1$? _____

> Answer: $P(1) = 3 - 2 + 2 + 1 = 4 \neq 0$; hence $x - 1$ is not a factor.
> $P(-1) = 3 - 2 - 2 + 1 = 0$; hence $x - (-1) = x + 1$ is a factor.

Problem Set 1.8

1. Find all roots of $2x^4 + x^3 + x^2 + 2x - 6 = 0$.

2. Find the remainder when $x^{17} - 3x^5 + 1$ is divided by $x + 1$.

3. Is $x - 1$ a factor of
   (a) $x^{43} - 1$?
   (b) $x^{43} + 1$?
   (c) $2x^{12} + 4x^7 - 3x^2 - 3$?

4. Find a polynomial equation of lowest degree whose roots include -1 and $2i$.

5. If c and d are integers, explain why 3/2 cannot be a root of $4x^5 + cx^3 - dx + 5 = 0$.

6. Explain why $x^4 + 7x^2 + x - 5 = 0$ must have an irrational root in the interval $(0, 1)$.

7. Given $\quad ax^6 - bx^4 + cx^3 + dx^2 - ex + f = 0$, where $a, b, c, d, e,$ and f are positive real numbers, what is
   (a) the maximum number of positive real roots this equation can have?
   (b) the maximum number of negative real roots this equation can have?

Solutions and Comments for Problem Set 1.8

1. Checking possible rational roots, we find $P(1) = P(-\frac{3}{2}) = 0$; hence $P(x) = (x - 1)(x + \frac{3}{2})(x^2 + 2)$ and the roots are $1, -3/2, \sqrt{2}i,$ and $-\sqrt{2}i$.

   > Possible rational roots are $\pm 1, \pm 2, \pm 3,$ $\pm 6, \pm 1/2, \pm 3/2$.

2. Using the Remainder Theorem, the remainder is $(-1)^{17} - 3(-1)^5 + 1 = 3$.

3. (a) yes;
   (b) no;
   (c) yes.

   > In (b), if $P(x) = x^{43} + 1$, then $P(1) = 2 \neq 0$ $\Rightarrow x - 1$ is not a factor.

4. Roots must be $-1, 2i,$ and $-2i$. Hence an equation satisfying the stated conditions is $(x + 1)(x - 2i)(x + 2i) = 0$, or, upon simplification, $x^3 + x^2 + 4x + 4 = 0$.

> Note that a polynomial equation is requested and hence the root $2i$ must have a conjugate pair root, $-2i$. A common error is to think that $(x + 1)(x - 2i) = 0$ satisfies the stated conditions. This is *not* a polynomial equation since not all of its coefficients are real numbers.

5. Since 3 is not a factor of 5, 3/2 cannot be a root of the given polynomial equation (Rational Root Theorem).

6. If $P(x) = x^4 + 7x^2 + x - 5$, then $P(0) = -5$ and $P(1) = 4$. Hence $y = P(x)$ has a graph that crosses the x-axis in the interval $(0, 1)$; that is, $P(x) = 0$ has a root in $(0, 1)$. Using the Rational Root Theorem, the only possible rational roots are $1, -1, 5,$ and -5. Since none of these satisfy the equation, and none are in $(0, 1)$, the root in the interval must be irrational.

7. (a) $P(x) = ax^6 - bx^4 + cx^3 + dx^2 - ex + f$.

There are 4 sign changes; hence there are at most 4 positive real roots.

(b) $P(-x) = ax^6 - bx^4 - cx^3 + dx^2 + ex + f$.
There are 2 sign changes; hence there are at most 2 negative real roots.

> $P(x) = 0$ could have no positive real roots and no negative real roots; that is, it could possibly have six imaginary roots.

*Section 1.8A: Graphs in Polar Coordinates

Points can be represented by *polar coordinates* (r, θ), where r is the directed distance from the origin to the given point and θ is the angle with center at $(0, 0)$ and whose sides are the positive x-axis and the ray joining the origin to the point. If (x, y) represents the rectangular coordinates of the point, then

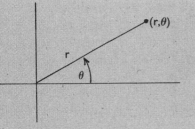

$$x = r \cos \theta \qquad r^2 = x^2 + y^2$$

$$y = r \sin \theta \qquad \tan \theta = \frac{y}{x}$$

Angles measured counterclockwise are positive. If r is negative, the distance $|r|$ is measured along the extension of the terminal side of θ through the origin.

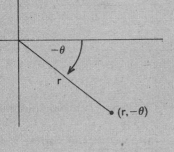

Authors' Note: A polar representation of a point is not unique. The point (r, θ) can be represented by $(r, \theta \pm 2\pi n)$ for $n = 1, 2, 3, \ldots$.

Q1. Convert to rectangular coordinates (Cartesian).

(a) $\left(2, \dfrac{2\pi}{3}\right)$ _____

(b) $\left(4, \dfrac{-\pi}{4}\right)$ _____

> Answers: (a) $\left(2 \cos \dfrac{2\pi}{3}, 2 \sin \dfrac{2\pi}{3}\right) = (-1, \sqrt{3})$.
>
> (b) $\left(4 \cos\left(-\dfrac{\pi}{4}\right), 4 \sin\left(-\dfrac{\pi}{4}\right)\right) = (2\sqrt{2}, -2\sqrt{2})$.

Q2. Convert to polar coordinates: (a) $(-2, -2\sqrt{3})$ (b) $(0, -2)$

> **Answers:** (a) $r = \sqrt{(-2)^2 + (-2\sqrt{3})^2} = \sqrt{16} = 4.$ $\operatorname{Tan}\theta = \dfrac{-2\sqrt{3}}{-2} = \sqrt{3} \Rightarrow \theta = \dfrac{\pi}{3}$ (Quad. I).
>
> Since the given point is in Quadrant III, polar coordinate representations could be $(-4, \pi/3)$ or $(4, 4\pi/3)$.
>
> (b) $r = \sqrt{(0)^2 + (-2)^2} = \sqrt{4} = 2.$ Since the given point is on the negative y-axis, a polar representation would be $(2, 3\pi/2)$.

Authors' Notes: In graphing, it is often easier to select values for special angles such as $30°, 45°, 60°, 90°,$ $120°,$ etc. and then find the value of r associated with these angles.

A polar graph is symmetric

    (i) about the polar axis (x-axis) if $f(\theta) = f(-\theta)$; that is, if θ and $-\theta$ both produce the same value for r.

    (ii) about the vertical line $\theta = \pi/2$ if $f(\theta) = f(\pi - \theta)$.

    (iii) about the pole (origin) if (r, θ) can be replaced by $(-r, -\theta)$ with no effect.

Determining symmetry (if it exists) reduces greatly the amount of point plotting necessary to determine a graph.

It is often useful to look at both polar and rectangular equations of a given curve since one may be considerably easier to graph than the other.

Q3. Discuss symmetry for: (a) $r = \sin\theta$ 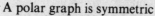   (b) $r^2 = \cos 2\theta$ _____

> **Answers:** (a) Since $r = \sin\theta$ and $-r = \sin(-\theta)$ represent the same set of points, the graph is symmetric about $\theta = \pi/2.$
>
> (b) Graph is symmetric about the polar axis, the pole, and the line $\theta = \pi/2.$

Authors' Note: $r = f(\theta)$ can be a function in polar coordinates, but not in rectangular coordinates. Part (a) in Q3 represents such a situation.

Q4. Identify the geometric figure described by

    (a) $r\cos\theta = 2$ _____   (b) $r = 4\cos\theta$ _____

> **Answers:** (a) A vertical line (rectangular equation is $x = 2$).
>
> (b) A circle. $\sqrt{x^2 + y^2} = \dfrac{4x}{\sqrt{x^2 + y^2}} \Rightarrow x^2 + y^2 = 4x \Rightarrow (x - 2)^2 + y^2 = 2^2.$ Center at $(2, 0),$ radius $= 2.$

Example: To sketch the graph of the cardioid $r = 1 + \sin\theta$, we would plot points (r, θ) to obtain the general shape. Some of the points on the graph are shown in the figure on the right.

r	θ
1	0
$1\frac{1}{2}$	$\frac{\pi}{6}$
2	$\frac{\pi}{2}$
1	π
$\frac{1}{2}$	$\frac{7\pi}{6}$
0	$\frac{3\pi}{2}$

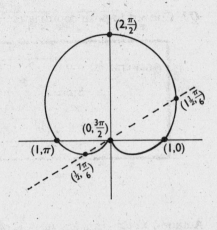

Problem Set 1.8A

1. Identify the geometric figure described by
   (a) $r = 3$
   (b) $\theta = -\pi/4$
   (c) $r = \sin\theta$
   (d) $r = \theta$

2. Write the polar equation for
   (a) the line $y = \sqrt{3}x$
   (b) $x^2 + y^2 = 36$

3. Convert $r^2 \cos 2\theta = 1$ to rectangular coordinates and identify the graph.

4. Sketch the graph of $r^2 = \cos 2\theta$.

5. Sketch the graph of
   (a) $r = 2\cos\theta$
   (b) $r = \cos 2\theta$
   (c) $r = \cos 3\theta$

6. Find the points of intersection for
   (a) $r = \sin\theta$ and $r = 1 - \sin\theta$
   (b) $r = 1 + \cos\theta$ and $r = 2\cos\theta$

Solutions and Comments for Problem Set 1.8A

1. (a) circle;
   (b) line;
   (c) circle;
   (d) spiral.

   > Circle in (c) has rectangular equation $x^2 + (y - \frac{1}{2})^2 = (\frac{1}{2})^2$.

2. (a) $\theta = \pi/3$;
   (b) $r = 6$.

   > Note the simplicity of the polar equations.

3. $r^2 \cos 2\theta = 1 \Rightarrow r^2(\cos^2\theta - \sin^2\theta) = 1 \Rightarrow$
   $r^2\cos^2\theta - r^2\sin^2\theta = 1 \Rightarrow$
   $(r\cos\theta)^2 - (r\sin\theta)^2 = 1 \Rightarrow x^2 - y^2 = 1.$
   Hence graph is a hyperbola.

   > In this instance, converting to rectangular coordinates would make graphing easier.

4.

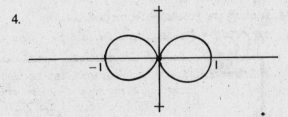

The graph is called a *lemniscate*. The rectangular equation is $x^4 + 2x^2y^2 + y^4 - x^2 + y^2 = 0$. It is clearly easier to graph in polar form.

In general, the graph of $r = \cos n\theta$ is an n-leaved rose if n is an odd integer and a $2n$-leaved rose if n is an even integer. A similar statement is true for $r = \sin n\theta$.

5.

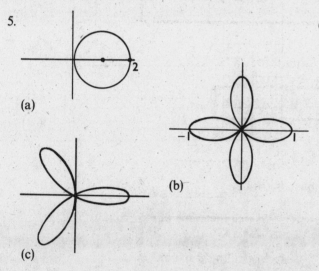

(a)

(b)

(c)

6. (a) $\sin\theta = 1 - \sin\theta \Rightarrow \sin\theta = \frac{1}{2} \Rightarrow \theta = \pi/6$, $5\pi/6$ if $\theta \in [0, 2\pi]$. The points of intersection are $(\frac{1}{2}, \frac{\pi}{6})$, $(\frac{1}{2}, \frac{5\pi}{6})$, and the origin.

(b) $1 + \cos\theta = 2\cos\theta \Rightarrow \cos\theta = 1$. The points of intersection are $(2, 0)$ and the origin.

There are, of course, other ways to represent the points of intersection. For instance, $(\frac{1}{2}, \frac{\pi}{6})$ is the same as $(\frac{1}{2}, \frac{13\pi}{6})$ or $(-\frac{1}{2}, \frac{7\pi}{6})$. Note carefully that in both (a) and (b), the origin is a point of intersection although it does not occur at the same angle in each equation. In (b), $(0, \pi)$ is a point on $r = 1 + \cos\theta$ and $(0, \frac{\pi}{2})$ is a point on $r = 2\cos\theta$.

Section 1.9: Trigonometric Identities and Addition Formulas

Basic identities:

(1) $\sin^2\theta + \cos^2\theta = 1$

(2) $\dfrac{\sin\theta}{\cos\theta} = \tan\theta$ if $\cos\theta \neq 0$

(3) $\csc\theta = \dfrac{1}{\sin\theta}$ if $\sin\theta \neq 0$

(4) $\sec\theta = \dfrac{1}{\cos\theta}$ if $\cos\theta \neq 0$

(5) $\cot\theta = \dfrac{1}{\tan\theta}$ if $\tan\theta \neq 0$

(6) $1 + \tan^2\theta = \sec^2\theta$

(7) $1 + \cot^2\theta = \csc^2\theta$

Sum and difference formulas:

(8) $\sin(\theta + \beta) = \sin\theta\cos\beta + \cos\theta\sin\beta$

(9) $\sin(\theta - \beta) = \sin\theta\cos\beta - \cos\theta\sin\beta$

(10) $\cos(\theta + \beta) = \cos\theta\cos\beta - \sin\theta\sin\beta$

(11) $\cos(\theta - \beta) = \cos\theta\cos\beta + \sin\theta\sin\beta$

(12) $\tan(\theta + \beta) = \dfrac{\tan\theta + \tan\beta}{1 - \tan\theta\tan\beta}$

(13) $\tan(\theta - \beta) = \dfrac{\tan\theta - \tan\beta}{1 + \tan\theta\tan\beta}$

Common error: $\sin(\theta + \beta) = \sin\theta + \sin\beta$.

Authors' Note: The above error should not be made by a "thinking" person. The left hand side of the equation cannot exceed 1, but the right hand side could be as large as 2 (if $\theta = \beta = \frac{\pi}{2}$, for instance). Other useful formulas related to the above appear in the Appendix.

Q1. Simplify (a) $\dfrac{\cos^2 x}{\sin x} + \sin x$ _____

(b) $(1 + \tan^2 \theta)\cos^2 \theta$ _____

Answers: (a) $\dfrac{\cos^2 x + \sin^2 x}{\sin x} = \dfrac{1}{\sin x} = \csc x.$ (b) $(\sec^2 \theta)(\cos^2 \theta) = 1.$

Q2. Writing $2x$ as $x + x$, use formula (8) to find a formula for $\sin 2x$. _____

Answer: $\sin(x + x) = \sin x \cos x + \cos x \sin x = 2 \sin x \cos x.$

Q3. Use sum and difference formulas to find: (a) $\sin 105°$ _____

(b) $\cos 15°$ _____   (c) $\tan \dfrac{5\pi}{12}$ _____

Answers: (a) $\sin(60° + 45°) = \sin 60° \cos 45° + \cos 60° \sin 45°$

$$= \frac{\sqrt{3}}{2} \times \frac{\sqrt{2}}{2} + \frac{1}{2} \times \frac{\sqrt{2}}{2} = \frac{\sqrt{6} + \sqrt{2}}{4}.$$

(b) $\cos(45° - 30°) = \cos 45° \cos 30° + \sin 45° \sin 30°$

$$= \frac{\sqrt{2}}{2} \times \frac{\sqrt{3}}{2} + \frac{\sqrt{2}}{2} \times \frac{1}{2} = \frac{\sqrt{6} + \sqrt{2}}{4}.$$

(c) $\tan\left(\dfrac{\pi}{4} + \dfrac{\pi}{6}\right) = \dfrac{\tan \frac{\pi}{4} + \tan \frac{\pi}{6}}{1 - \tan \frac{\pi}{4} \tan \frac{\pi}{6}} = \dfrac{1 + \frac{\sqrt{3}}{3}}{1 - 1(\frac{\sqrt{3}}{3})} = \dfrac{3 + \sqrt{3}}{3 - \sqrt{3}} = 2 + \sqrt{3}.$

Problem Set 1.9

1. Prove that $(1 - \sin^2 x)(1 + \tan^2 x) = 1.$

2. Prove that $\sin^4 x - \cos^4 x = \sin^2 x - \cos^2 x.$

3. Prove that $(1 - \tan x)^2 = \sec^2 x - 2\tan x.$

4. Use the sum and difference formulas to calculate

(a) $\sin \frac{\pi}{12}$
(b) $\cos 75°$
(c) $\tan 15°$

5. Prove
(a) $\sin(\pi + \theta) = -\sin \theta$
(b) $\sin(\frac{\pi}{2} + x) = \cos x$
(c) $\cos 2x = \cos^2 x - \sin^2 x$

Solutions and Comments for Problem Set 1.9

1. $(1 - \sin^2 x)(\sec^2 x) = \dfrac{1 - \sin^2 x}{\cos^2 x} = \dfrac{\cos^2 x}{\cos^2 x} = 1.$

$\sin^2 x + \cos^2 x = 1 \Rightarrow \cos^2 x = 1 - \sin^2 x.$

2. Factor the left side:
$(\sin^2 x + \cos^2 x)(\sin^2 x - \cos^2 x) =$
$\sin^2 x - \cos^2 x.$

$$a^4 - b^4 = (a^2 + b^2)(a^2 - b^2).$$

3. $(1 - \tan x)^2 = 1 - 2\tan x + \tan^2 x =$
$(1 + \tan^2 x) - 2\tan x = \sec^2 x - 2\tan x.$

$\sec^2 x = 1 + \tan^2 x$ is very useful in calculus.

4. (a) $\sin\left(\dfrac{\pi}{4} - \dfrac{\pi}{6}\right) = \dfrac{\sqrt{2}}{2} \times \dfrac{\sqrt{3}}{2} - \dfrac{\sqrt{2}}{2} \times \dfrac{1}{2}$
$$= \dfrac{\sqrt{6} - \sqrt{2}}{4};$$

   (b) $\cos(45° + 30°) = \dfrac{\sqrt{2}}{2} \times \dfrac{\sqrt{3}}{2} - \dfrac{\sqrt{2}}{2} \times \dfrac{1}{2}$
$$= \dfrac{\sqrt{6} - \sqrt{2}}{4};$$

   (c) $\tan(45° - 30°) = \dfrac{1 - \frac{\sqrt{3}}{3}}{1 + 1(\frac{\sqrt{3}}{3})} = \dfrac{3 - \sqrt{3}}{3 + \sqrt{3}}$
$$= 2 - \sqrt{3}.$$

The sine of an angle is equal to the cosine of its complement. Hence, it should not be surprising that answers (a) and (b) are equal.

Common error:
$\cos(\theta + \beta)$
$= \cos\theta\cos\beta + \sin\theta\sin\beta$

One can rationalize $\dfrac{3 - \sqrt{3}}{3 + \sqrt{3}}$ by multiplying numerator and denominator by $3 - \sqrt{3}$.

5. (a) $\sin\pi\cos\theta + \cos\pi\sin\theta$
$$= 0 \times \cos\theta + (-1)\sin\theta = -\sin\theta;$$
   (b) $\sin\frac{\pi}{2}\cos x + \cos\frac{\pi}{2}\sin x$
$$= 1 \times \cos x + 0 \times \sin x = \cos x;$$
   (c) $\cos(x + x) = \cos x\cos x - \sin x\sin x$
$$= \cos^2 x - \sin^2 x.$$

Common error:
$\sin(\frac{\pi}{2} + x) = \sin\frac{\pi}{2} + \sin x$
$$= 1 + \sin x$$

(c) is the famous double angle formula for $\cos 2x$. Note that $\cos 2x = \cos^2 x - \sin^2 x = 2\cos^2 x - 1 = 1 - 2\sin^2 x.$

Section 1.10: Amplitude and Periodicity of $A\sin(Bx + C)$ and $A\cos(Bx + C)$

Consider the functions

$$y = A\sin(Bx + C) \quad \text{and} \quad y = A\cos(Bx + C).$$

The *amplitude* of these periodic functions is one-half of the absolute value of the difference between the maximum and minimum values. In each case, the amplitude is $|A|$.

The *period* for these functions is $\dfrac{2\pi}{B}$.

Authors' Note: For both functions, we will obtain a complete graph of the sine or cosine curve as the angle $Bx + C$ assumes values from 0 to 2π. Assuming $B > 0$, we have $0 \le Bx + C \le 2\pi \Rightarrow$ $-C \le Bx < 2\pi - C \Rightarrow -\dfrac{C}{B} \le x \le \dfrac{2\pi - C}{B}$. Note that $\dfrac{2\pi - C}{B} - \left(-\dfrac{C}{B}\right) = \dfrac{2\pi}{B}$.

Q1. Find the amplitude and period for each function and sketch the graph on $[0, 2\pi]$.

(a) $y = 2\sin x$

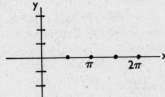

(b) $y = \sin 2x$

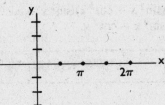

(c) $y = \cos\frac{1}{2}x$

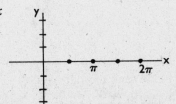

(d) $y = -3\cos x$

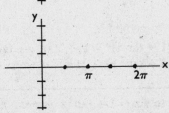

Answers: (a) $|A| = 2$, period $= 2\pi$.

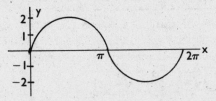

(b) $A = 1$, period $= \dfrac{2\pi}{2} = \pi$.

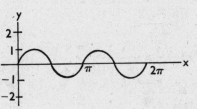

(c) $A = 1$, period $= \dfrac{2\pi}{\frac{1}{2}} = 4\pi$.

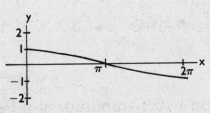

(d) $A = 3$, period $= 2\pi$.

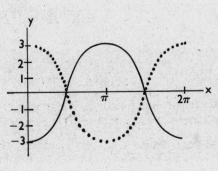

Authors' Note: For (d), a good strategy is to lightly sketch the graph of $y = 3\cos x$ (dotted in diagram) and then "flip" it around the x-axis to get the graph of $y = -3\cos x$.

The *phase shift* is the horizontal translation of the curve. For $y = A\sin(Bx + C)$ and $y = A\cos(Bx + C)$, the phase shift is equal to $\dfrac{C}{B}$.

Authors' Notes: It is sometimes helpful to write $y = A\sin(Bx + C)$ as $y = A\sin\left[B\left(x + \dfrac{C}{B}\right)\right]$.

If $\dfrac{C}{B} < 0$, the graph is shifted to the right.

If $\dfrac{C}{B} > 0$, the graph is shifted to the left.

Example: To sketch the graph of $y = 3\cos(2x + \frac{\pi}{2}) = 3\cos\left[2(x + \frac{\pi}{4})\right]$, we can note that amplitude $= 3$, period $= \frac{2\pi}{2} = \pi$, and the phase shift is $\frac{\pi}{4}$. The graph is shown below.

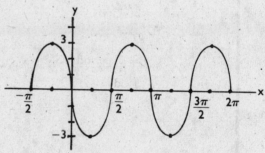

Authors' Notes: Students tend to make graphing of these functions much more difficult than necessary. Once the amplitude, period and phase shift have been determined, the graphing process should be quite easy. In the example above, we need only draw the cosine curve over an interval π units long, within the vertical range -3, 3, and starting at $-\frac{\pi}{4}$. Note that one usually "starts" a graph of the cosine curves when the angle is 0. In the example, $2x + \frac{\pi}{2} = 0$ when $x = -\frac{\pi}{4}$. Also note that if we had to graph $y = -3\cos(2x + \frac{\pi}{2})$, we could take the graph above and just "flip" it around the x-axis.

Problem Set 1.10

In problems 1–4, identify the amplitude, period, and phase shift.

1. $y = -2\sin(-2x)$

2. $y = -\pi\cos(\frac{\pi}{2}x + \pi)$

3. $y = \frac{1}{4}\cos(4x + 12)$

4. $y = \frac{3}{2}\sin(\frac{1}{2}x - 3)$

In problems 5–8, sketch the graph.

5. $y = 2\sin(\pi x + \pi)$ on $[0, 4]$

6. $y = -3\cos\left[2(x - 1)\right]$ over one period

7. $y = \frac{1}{2}\sin(\frac{1}{2}x + 2\pi)$ over one period

8. $y = 2 + \cos(x - \frac{\pi}{2})$ over one period

Solutions and Comments for Problem Set 1.10

	Amplitude	Period	Phase Shift
1.	2	π	None
2.	π	4	2 units to the left. $\frac{\pi}{2}x + \pi = 0$ when $x = -2$.
3.	$\frac{1}{4}$	$\frac{\pi}{2}$	3 units to the left. $4x + 12 = 0$ when $x = -3$.
4.	$\frac{3}{2}$	4π	6 units to the right. $\frac{1}{2}x - 3 = 0$ when $x = 6$.

We will use this area to emphasize that graphing functions of this type is not difficult. We will go through the steps of constructing the graph of $y = -\pi\cos(\frac{\pi}{2}x + \pi)$ in problem #2. Lightly sketch the lines $y = \pi$ and $y = -\pi$. (Amplitude $= \pi$)

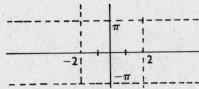

Lightly sketch the line $x = -2$, then move 4 units to the right and sketch $x = 2$. (Period $= 4$)

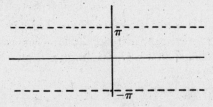

Now, lightly sketch the cosine curve in the rectangle. We now have one period of the function $y = \pi\cos(\frac{\pi}{2}x + \pi)$.

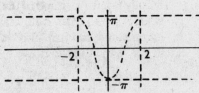

Finally, flip the graph around the x-axis. This produces one period of the graph of $y = -\pi\cos(\frac{\pi}{2}x + \pi)$.

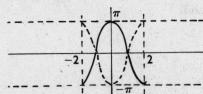

5.

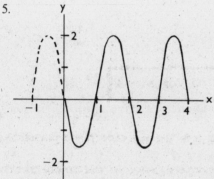

7.

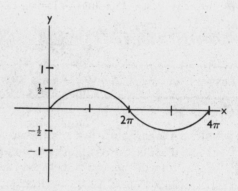

6.

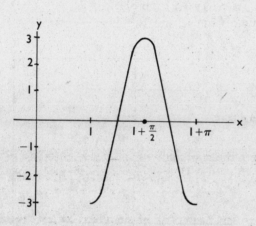

8.

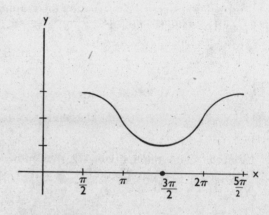

> This is just the graph of $y = \cos\left(x - \frac{\pi}{2}\right)$ shifted up two units. Just center the cosine curve on the line $y = 2$.

Section 1.11: Exponential and Logarithmic Functions

If $b > 0$, $b \neq 1$, and $x > 0$, then $y = \log_b x \Leftrightarrow b^y = x$. Basic laws of logarithms:

(1) $\log_b (ac) = \log_b a + \log_b c$

(2) $\log_b \left(\dfrac{a}{c}\right) = \log_b a - \log_b c$

(3) $\log_b a^c = c \log_b a$

(4) $\log_b \sqrt[n]{a} = \dfrac{1}{n} \log_b a$

(5) $\log_b a = \dfrac{\log_c a}{\log_c b}$

$Q1$. Evaluate: (a) $\log_{10} 100$ _____

(b) $\log_8 2$ _____

(c) $\log_2 \sqrt[3]{64}$ _____

(d) $\log_3 \left(\frac{81}{27}\right)$ _____

> Answers: (a) 2.
> (c) $\frac{1}{3} \log_2 64 = \frac{1}{3}(6) = 2$.
> (b) $8^x = 2 \Rightarrow (2^3)^x = 2 \Rightarrow 2^{3x} = 2 \Rightarrow 3x = 1 \Rightarrow x = 1/3$.
> (d) $\log_3 81 - \log_3 27 = 4 - 3 = 1$.

Q2. Solve for x: (a) $\log_{10} 25 + \log_{10} 40 = x$ _____

(b) $\dfrac{\log_{12} 64}{\log_{12} 4} = x$ _____

Answers: (a) $\log_{10}(25 \times 40) = \log_{10} 1000 = x \Rightarrow x = 3$; (b) $\log_4 64 = 3$.

 Authors' Notes: A logarithm is just an exponent. The expression $\log_b x$ is "the exponent (power) to which b must be raised in order to obtain x."
A logarithm can be negative (there are negative exponents); however, you cannot take the logarithm of a negative number since the domain of $y = \log_b x$ is $(0, \infty)$.

 Common errors: $\dfrac{\log_b a}{\log_b c} = \log_b a - \log_b c$.

$$\log_b\left(\dfrac{a}{c}\right) = \dfrac{\log_b a}{\log_b c}.$$

$$\log_b(ac) = \log_b a \log_b c.$$

$$(\log_b a)(\log_b c) = \log_b a + \log_b c.$$

The functions $y = \log_b x$ and $y = b^x$ are inverses. Note that if $f(x) = \log_b x$ and $g(x) = b^x$, then $(f \circ g)(x) = f(b^x) = \log_b b^x = x \log_b b = x$ and $(g \circ f)(x) = g(\log_b x) = b^{\log_b x} = x$. Hence $g = f^{-1}$ and $f = g^{-1}$.

 Authors' Note: A complicated looking expression such as $b^{\log_b x}$ can often be simplified by just "reading" it. The expression $b^{\log_b x}$ is "b raised to the power to which b must be raised in order to get x." If you get anything other than x from this expression, you should be surprised . . . and you didn't read carefully.

Q3. On the axes provided, sketch the graphs of $y = \log_2 x$ and $y = 2^x$.

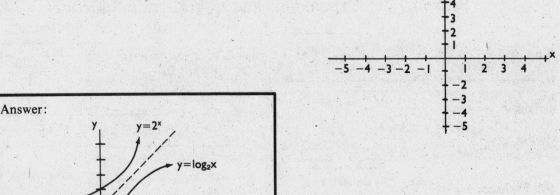

Answer:

Authors' Note: The dotted line on the graph is $y = x$. Remember that the graphs of a function and its inverse are symmetric with respect to the line $y = x$.

Problem Set 1.11

1. Write in logarithmic notation:
   (a) $8^{(2/3)} = 4$
   (b) $2^{-2} = \frac{1}{4}$

2. Solve for x: $\log_5 x < 0$.

3. Evaluate
   (a) $\log_3 27$
   (b) $\log_{125}\left(\frac{1}{5}\right)$
   (c) $\log_w w^{45}$

4. If $\log_b a = 5$ and $\log_b c = 2$, evaluate
   (a) $(\log_b a)(\log_b c)$
   (b) $\log_b(ac)$

   (c) $\log_b\left(\dfrac{c}{a}\right)$
   (d) $\dfrac{\log_b c}{\log_b a}$
   (e) $\log_b a^{\log_b c}$
   (f) $\log_a b$
   (g) $\log_c b$

5. In each instance, find $f^{-1}(x)$:
   (a) $f(x) = 7^x$
   (b) $f(x) = \log_{31} x$

6. Prove that $\log_b a = \dfrac{1}{\log_a b}$.

Solutions and Comments for Problem Set 1.11

1. (a) $\log_8 4 = 2/3$;
   (b) $\log_2\left(\frac{1}{4}\right) = -2$.

   > A log can be negative, as in (b), but you can't take the log of a negative number.

2. $x \in (0, 1)$.

   > If $0 < x < 1$, then $\log_b x < 0$. Note that if $x = 1$, $\log_b 1 = 0$.

3. (a) 3;
   (b) $125^x = \frac{1}{5} \Rightarrow 5^{3x} = 5^{-1} \Rightarrow x = -1/3$;
   (c) $45 \log_w w = 45$.

   > Read (c) . . . "the exponent to which w must be raised in order to obtain w^{45}."

4. (a) $5 \times 2 = 10$;
   (b) $5 + 2 = 7$;
   (c) $2 - 5 = -3$;

   (d) $2/5$;
   (e) $2\log_b a = 2 \times 5 = 10$;
   (f) $\log_b a = 5 \Rightarrow b^5 = a \Rightarrow b = a^{1/5}$
       $\Rightarrow \log_a b = 1/5$;
   (g) $1/2$.

   > Don't confuse (a) and (b) . . . or (c) and (d). Parts (f) and (g) relate to an important relationship referenced in problem #6.

5. (a) $f^{-1}(x) = \log_7 x$;
   (b) $f^{-1}(x) = 31^x$.

   > Check by calculating $f \circ f^{-1}$ and $f^{-1} \circ f$.

6. Let $\log_b a = x$ and $\log_a b = y$. Then $b^x = a$ and $a^y = b \Rightarrow (a^y)^x = a \Rightarrow a^{xy} = a \Rightarrow xy = 1 \Rightarrow x = 1/y$.

   > Relate this to pairs (f) and (g) in problem #4.

Section 1.12: Multiple Choice Questions for Chapter 1

1. A neighborhood of -4 with radius 2 is described by
   (A) $|x - 4| \leq 2$
   (B) $|x - 4| < 2$
   (C) $|x - 2| < 4$
   (D) $|x| < 2$
   (E) $|x + 4| < 2$

2. If $f(x) = 2x - 8$, then $f^{-1}(t) =$
   (A) $\dfrac{1}{2t - 8}$
   (B) $\dfrac{t + 8}{2}$
   (C) $\dfrac{t}{2} + 8$
   (D) $2t + 8$
   (E) $\frac{1}{2}t - 8$

3. If $c(x) = \cos(\sin^{-1} x)$, what is the range of c?
   (A) $[0, \frac{\pi}{2}]$
   (B) $[-\frac{\pi}{2}, \frac{\pi}{2}]$
   (C) $[-1, 1]$
   (D) $[0, 1]$
   (E) $[-1, 0]$

4. If $6x^7 - ax^5 + bx^3 - cx - 15 = 0$, where a, b, and c are integers, which of the following could *not* be a root of the polynomial equation?
   (A) $3/2$
   (B) $3/4$
   (C) $1/6$
   (D) $5/3$
   (E) -15

5. $\dfrac{1}{1 + \sin x} + \dfrac{1}{1 - \sin x} =$
   (A) $\dfrac{1}{1 - \sin^2 x}$
   (B) $2 \sec^2 x$
   (C) $\cos x$
   (D) $\tan^2 x$
   (E) $\csc^2 x$

6. The graph of $|y| = |x|$ is
   (A) a circle
   (B) a line
   (C) two intersecting lines
   (D) a hyperbola
   (E) a single point

7. If $f(x) = x^5 - 3x^2 + 2x - 7$ and $g(x) = 3x$, then $g(f(x)) =$
   (A) $x^6 - 9x^3 + 2x^2 - 7x$
   (B) $3x^5 - 9x^2 + 6x - 21$
   (C) $3x$
   (D) $243x^5 - 27x^3 + 6x - 21$
   (E) $3x^6 - 9x^3 + 6x^2 - 21x$

8. *How many* of the following functions have the property $f(x) = f(-x)$?
   (i) $f(x) = x^4 + x^2$
   (ii) $f(x) = \cos x$
   (iii) $f(x) = 4^x$
   (iv) $f(x) = 17$
   (A) 0
   (B) 1
   (C) 2
   (D) 3
   (E) 4

9. $\sin^2 x - \cos^2 x =$
   (A) 1
   (B) $2 \sin^2 x - 1$
   (C) $(\sin x - \cos x)^2$
   (D) $2 \cos^2 x - 1$
   (E) $\tan^2 x$

10. Solve for x: $2x = 7^{1 + \log_7 4}$.
    (A) 3
    (B) 6
    (C) 7
    (D) 10
    (E) 14

11. What is the period of the function
    $$y = 2 \cos\left(\frac{x}{3} - \frac{1}{2}\right)?$$
    (A) 2
    (B) $\pi/3$
    (C) 3
    (D) 6π
    (E) π

12. If $f(x) = x^2 + 1$ and $g(x) = x + 1$, then the solution set of $f(g(x)) = g(f(x))$ is
    (A) $\{0, 1\}$
    (B) $\{0\}$
    (C) all reals
    (D) $\{x: x \geq 0\}$
    (E) $\varnothing$

13. Which graph represents the solution set of $|x + 1| + |x - 1| = 2x$?

    (A)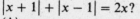

    (B)

    (C)

    (D)

    (E)

14. If $f(x) = x^5 + x^3 + 1$ and $f(a) = 10$, then $f(-a) =$
    (A) -7
    (B) 10
    (C) -10
    (D) 8
    (E) -8

15. An asymptote for $y = \dfrac{(x + 2)(x - 7)}{x - 5}$ is

    (A) $x = 0$
    (B) $x = -2$
    (C) $x = 5$
    (D) $x = -5$
    (E) $y = -2$

16. Which one of the graphs represents $y = \sin(2x - \pi)$?

    (A)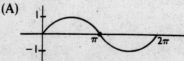

    (B)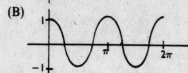

    (C)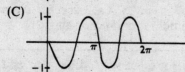

    (D)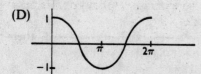

    (E)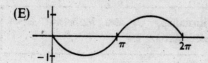

17. $\log_{1/b} x =$
    (A) $-\log_b x$
    (B) $\log_x b$
    (C) $-\log_x b$
    (D) b^x
    (E) none of the answer choices is correct.

18. Which of the following is a factor of $x^{17} - 4x^{15} - x^3 + 4$?
    (A) $x + 2$
    (B) $x - 2$
    (C) $x + 1$
    (D) $x - 1$
    (E) $x - 3$

19. If f and g are odd and even functions, respectively, such that $f(a) = b$ and $g(c) = b$, $b \neq 0$, then $\dfrac{f(-a)}{g(-c)} + f(-a) - g(-c) =$
    (A) -1
    (B) $-1 - 2b$
    (C) $-1 + 2b$
    (D) $1 - 2b$
    (E) $1 + 2b$

20. The set of all points $\{(10^t, t)\}$ is described by the equation
    (A) $y = 10^x$
    (B) $x = 10y$
    (C) $x = \log_{10} y$
    (D) $y = \log_{10} x$
    (E) $\log_{10} x = \log_{10} y$

21. Which of the following functions is periodic with period π?
    (A) $y = \sin x$
    (B) $y = \pi x$
    (C) $y = |\cos x|$
    (D) $y = \pi$
    (E) $y = \sin \pi x$

22. If $f(x) = \log_b x$ and $g(x) = b^x$, then $f(g(x)) =$
    (A) 1
    (B) x
    (C) x^b
    (D) b^x
    (E) $\log_x b$

23. $f(a + b) = f(a) \times f(b)$ is true if $f(x) =$
    (A) x
    (B) $\log_c x$
    (C) $\sin x$

(D) $1/x$

(E) 3^x

24. *How many* of the following functions are symmetric about the y-axis?

  (i) $y = x^2$

  (ii) $y = \sqrt{4 - x^2}$

  (iii) $y = x$

  (iv) $y = \cos x$

(A) 0

(B) 1

(C) 2

(D) 3

(E) 4

25. If $w(x) = 5x + 10$ and $z(x) = -x^2 + 6$, then $(w \circ z \circ w^{-1})(0) =$

(A) 20

(B) -10

(C) 15

(D) 6

(E) 12

*26. The graph described by $x = 4\cos t$, $y = 4\sin t$ is

(A) a line

(B) a parabola

(C) a circle

(D) an ellipse

(E) none of the answers is correct.

*27. A parametric representation for the line containing $(-3, 6)$ and $(2, 10)$ is

(A) $x = -3 + 4t$

  $y = \quad 6 + 5t$

(B) $x = 4 + t$

  $y = 5 + t$

(C) $x = -3 + 2t$

  $y = \quad 6 + 3t$

(D) $x = 1 - 3t$

  $y = 1 + 6t$

(E) $x = \quad 2 + 5t$

  $y = 10 + 4t$

*28. The polar coordinates of P are $(-5, \frac{\pi}{2})$. What are the rectangular coordinates?

(A) $(0, 5)$

(B) $(5, 0)$

(C) $(2, 2)$

(D) $(0, -5)$

(E) $(-5, 0)$

*29. Which of the following represents the shape of the graph of the polar equation $r = 2$?

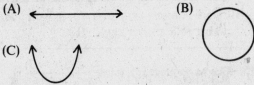

(A)   (B)

(C)

   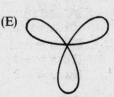

(D)   (E)

*30. If $r = 4\sin\theta$ is converted to rectangular coordinates, then

(A) $x^2 + (y - 2)^2 = 4$

(B) $(x - 2)^2 + y^2 = 4$

(C) $x^2 + y^2 = 16$

(D) $x^2 + 4y^2 = 4$

(E) $(x - 2)^2 + (y - 2)^2 = 4$

Solutions and Comments for Multiple Choice Questions

1. **(E)**

> (A) or (B) would be very careless choices.

2. **(B)** f is a rule that says "take a number, multiply it by 2, then subtract 8." Hence, f^{-1} has to be a rule that says "take a number, add 8, then divide by 2."

> Don't confuse $f^{-1}(t)$ with $[f(t)]^{-1}$. Note that (A) is $[f(t)]^{-1}$.

3. **(D)** $-\pi/2 \le \sin^{-1} x \le \pi/2$ and the cosine of angles in this interval is non-negative.

> Since $\cos x \le 1$, answer choices (A) and (B) can be eliminated from consideration immediately.

4. **(B)**

> An application of the Rational Root Theorem, Section 1.8.

5. **(B)** $\dfrac{(1 - \sin x) + (1 + \sin x)}{1 - \sin^2 x} = \dfrac{2}{\cos^2 x}$

$= 2 \sec^2 x.$

$$\boxed{\cos^2 x + \sin^2 x = 1 \Rightarrow \cos^2 x = 1 - \sin^2 x.}$$

6. **(C)** $y = x$ or $y = -x.$

$$\boxed{\text{The absolute value of a number is either the number or its opposite.}}$$

7. **(B)** Rule g merely says "take a number and multiply it by 3."

$$\boxed{\text{If you confuse } f \circ g \text{ and } f \times g, \text{ you can end up with answer choice (E).}}$$

8. **(D)** $f(x) = 4^x$ does not have the property.

$$\boxed{\text{You are looking for } \textit{even} \text{ functions.}}$$

9. **(B)** $\sin^2 x - (1 - \sin 2x) = 2 \sin^2 x - 1.$

$$\boxed{\text{Note that } \sin^2 x - \cos^2 x = -\cos 2x.}$$

10. **(E)** $2x = 7 \times 7^{\log_7 4} = 7 \times 4 = 28 \Rightarrow x = 14.$

$$\boxed{7^{\log_7 4} \text{ is "seven raised to the power to which 7 must be raised in order to obtain 4."}}$$

11. **(D)** $\dfrac{2\pi}{1/3} = 6\pi.$

$$\boxed{y = 2\cos\left[\tfrac{1}{3}\left(x - \tfrac{3}{2}\right)\right].}$$

12. **(B)** $f(g(x)) = (x + 1)^2 + 1$ and $g(f(x)) = (x^2 + 1) + 1.$ $x^2 + 2x + 2 = x^2 + 2 \Rightarrow x = 0.$

13. **(D)** If $x \geq 1$, then $x + 1 + x - 1 = 2x \Rightarrow 2x = 2x.$ Hence $x \geq 1$ is part of a solution.

If $-1 < x < 1$, then $x + 1 + 1 - x = 2x \Rightarrow 2 = 2x \Rightarrow x = 1.$ Hence $(-1, 1)$ contains no numbers that satisfy the given equation.

If $x \leq -1$, then $-x - 1 + 1 - x = 2x \Rightarrow -2x = 2x \Rightarrow x = 0.$ Hence $(-\infty, -1]$ contains no numbers that satisfy the given equation.

$$\boxed{\text{Answer choice (E) can be eliminated from consideration immediately by noting that if } x < 0, \text{ then } 2x \text{ is negative and } |x + 1| + |x - 1| \text{ is non-negative.}}$$

14. **(E)** $f(a) = 10$ $a^5 + a^3 = 9.$ Hence
$f(-a) = -a^5 - a^3 + 1 = -(a^5 + a^3) + 1$
$= -9 + 1 = -8.$

$$\boxed{f \text{ is not an odd function. Note that } f(x) = x^5 + x^3 + 1x^0.}$$

15. **(C)**

$$\boxed{\text{Function is undefined at } x = 5.}$$

16. **(C)** We need a complete cycle of the sine curve starting at $x = \frac{\pi}{2}$ with period $\frac{2\pi}{2} = \pi.$

$$\boxed{\text{Note that if } x = 0, \text{ then } y = \sin(-\pi) = 0. \text{ This immediately eliminates answer choices (B) and (D).}}$$

17. **(A)** $\left(\dfrac{1}{b}\right)^y = x \Rightarrow b^y = x^{-1}$
$\Rightarrow y = \log_b x^{-1} = -\log_b x.$

18. **(D)** If $x = 1$, then the value of the polynomial is 0. Factor Theorem (Section 1.8) $\Rightarrow x - 1$ is a factor.

19. **(B)** $\dfrac{-b}{b} + (-b) - b = -1 - 2b.$

$$\boxed{f(-a) = -b \text{ and } g(-c) = b.}$$

20. **(D)** $x = 10^y \Rightarrow y = \log_{10} x$.

> Answer choice (A) is a careless response.

21. **(C)** Note the graph of $y = |\cos x|$.

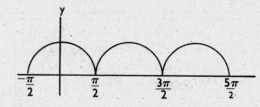

> Easy to get sloppy and choose (E). The period for (E) is $\frac{\pi}{\pi} = 1$.

22. **(B)** The functions are inverses of each other.

23. **(E)** If $f(x) = 3^x$, then
$f(a + b) = 3^{a+b} = 3^a \times 3^b = f(a) \times f(b)$.

24. **(D)**

> Function (ii) is a semicircle. It is a subset of $x^2 + y^2 = 4$.

25. **(A)** $w^{-1}(x) = \dfrac{x - 10}{5}$; hence $(w \circ z)(w^{-1}(0)) =$
$(w \circ z)(-2) = w(z(-2)) = w(2) = 20$.

> Apply rule w^{-1} first, then rule z, finally rule w.

26. **(C)** $x^2 + y^2 = 16(\cos^2 t + \sin^2 t) = 16$.
In general, $x = r \cos t$, $y = r \sin t$ represents a circle with center at $(0, 0)$ and radius r.

27. **(E)** The direction vector is $(2 - (-3), 10 - 6)$
$= (5, 4)$. Another representation would be
$x = -3 + 5t$, $y = 6 + 4t$.

28. **(D)** $x = -5 \cos \frac{\pi}{2} = 0$ and
$y = -5 \sin \frac{\pi}{2} = -5$.

29. **(B)** The polar equation $r = 2$ is merely describing all points whose distance from $(0, 0)$ is 2 units.

30. **(A)** $\sqrt{x^2 + y^2} = \dfrac{4y}{\sqrt{x^2 + y^2}}$
$\Rightarrow x^2 + y^2 - 4y = 0$
$\Rightarrow x^2 + y^2 - 4y + 4 = 4$
$\Rightarrow x^2 + (y - 2)^2 = 4$.

Chapter 2
LIMITS

Section 2.1: Intuitive Definition and Properties of Limits

> Definition (intuitive): The statement
> $$\lim_{x \to a} f(x) = L$$
> means that if x is close to and different from a, then $f(x)$ is close to and may equal L.

$Q1$. Evaluate: (a) $\lim_{x \to 3} x^2$ _____   (b) $\lim_{x \to 27} 5$ _____

(c) $\lim_{w \to 0} \dfrac{1}{w}$ _____   (d) $\lim_{h \to 0} \dfrac{4(x+h) - 4x}{h}$ _____

Answers: (a) 9.   (b) 5.   (c) Limit does not exist.   (d) $\lim_{h \to 0} \dfrac{4h}{h} = \lim_{h \to 0} 4 = 4.$

Authors' Note: In (c), $\dfrac{1}{w}$ gets infinitely large as w approaches 0. One can write $\lim_{w \to 0} \dfrac{1}{w} = \infty$. This basically says that $\dfrac{1}{w}$ gets larger and larger and does not approach any specific number as x gets closer and closer to 0.

Common error: $\lim_{x \to 0} \dfrac{4x}{x}$ doesn't exist because $\dfrac{4 \times 0}{0} = \dfrac{0}{0}$ is indeterminate.

Authors' Note: If $x \neq 0$, then $\dfrac{4x}{x} = 4$. Remember that $\lim_{x \to 0} \dfrac{4x}{x}$ is the number (if it exists) that $\dfrac{4x}{x}$ approaches when x gets close to, but not equal to, 0. The graph of $y = \dfrac{4x}{x}$ is shown at the top of page 49. Note the deleted point.

48

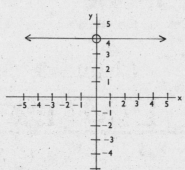

If $\lim\limits_{x \to a} f(x) = L_1$ and $\lim\limits_{x \to a} g(x) = L_2$, then

(1) $\lim\limits_{x \to a} [f(x) + g(x)] = L_1 + L_2$

(2) $\lim\limits_{x \to a} [f(x) - g(x)] = L_1 - L_2$

(3) $\lim\limits_{x \to a} [f(x) \times g(x)] = L_1 \times L_2$

(4) $\lim\limits_{x \to a} \dfrac{f(x)}{g(x)} = \dfrac{L_1}{L_2}$ if $L_2 \neq 0$

Q2. Evaluate: (a) $\lim\limits_{x \to 1} (x^2 + 4x - 2)$ _____

(b) $\lim\limits_{x \to \pi/2} (x^2 \sin x)$ _____

(c) $\lim\limits_{x \to 100} \dfrac{\log_{10} x}{x}$ _____

(d) $\lim\limits_{x \to 2} \dfrac{x^2 - 4}{x - 2}$ _____

Answers: (a) $1 + 4 - 2 = 3.$ (b) $\left(\dfrac{\pi}{2}\right)^2 \times 1 = \dfrac{\pi^2}{4}.$

(c) $\dfrac{2}{100} = \dfrac{1}{50}.$ (d) $\lim\limits_{x \to 2} (x + 2) = 4.$

 Common error: Applying rule (4) to $\lim\limits_{x \to 2} \dfrac{x^2 - 4}{x - 2}$, one gets $\dfrac{\lim\limits_{x \to 2}(x^2 - 4)}{\lim\limits_{x \to 2}(x - 2)} = \dfrac{0}{0}$, which is indeterminate.

Authors' Note: Rule (4) cannot be applied in (d), since $\lim\limits_{x \to 2} (x - 2) = 0$. (See the restriction in the rule statement.) However, the limit in (d) still exists. If one can't apply a limit rule, this does not mean that the limit in question does not exist.

Q3. Evaluate: (a) $\lim\limits_{x \to 0} \dfrac{x^2}{x}$ _____

(b) $\lim\limits_{x \to 0} \dfrac{x}{x^2}$ _____

(c) $\lim\limits_{x \to 0} \dfrac{x}{x}$ _____

Answers: (a) $\lim\limits_{x \to 0} x = 0.$ (b) $\lim\limits_{x \to 0} \dfrac{1}{x} = \infty.$ (c) $\lim\limits_{x \to 0} 1 = 1.$

Authors' Note: In Q3, one gets $\dfrac{0}{0}$ by actually replacing x by 0 in each case. However, each function "behaves" differently from the other two as x approaches 0.

Problem Set 2.1

In problems 1–9, find each limit if it exists.

1. $\lim\limits_{x \to 2} (x^2 - 3x + 4)$

2. $\lim\limits_{x \to \pi/2} \dfrac{\sin x}{x}$

3. $\lim\limits_{x \to 3} \dfrac{x}{x^3 + 2}$

4. $\lim\limits_{w \to -5} \dfrac{w^2 - 25}{w + 5}$

5. $\lim\limits_{x \to 1} \log_8 x$

6. $\lim\limits_{w \to 2.5} \dfrac{2}{w - 2}$

7. $\lim\limits_{t \to 2} \dfrac{2}{t - 2}$

8. $\lim\limits_{x \to 2} \dfrac{x^2 - 3x + 2}{x^2 - x - 2}$

9. $\lim\limits_{h \to 0} \dfrac{(x + h)^2 - x^2}{h}$

10. Find $\lim\limits_{x \to 1} \dfrac{x^3 - x^2}{x - 1}$ and sketch the graph of $y = \dfrac{x^3 - x^2}{x - 1}$.

11. Sketch the graph of $f(x) = \begin{cases} x + 2 & \text{if } x \geq 0 \\ x - 2 & \text{if } x < 0 \end{cases}$.
Does $\lim\limits_{x \to 0} f(x)$ exist?

Solutions and Comments for Problem Set 2.1

1. $2^2 - 3 \times 2 + 4 = 2$.

2. $\dfrac{1}{\pi/2} = \dfrac{2}{\pi}$.

3. $\dfrac{3}{3^3 + 2} = \dfrac{3}{29}$.

4. $\lim\limits_{w \to -5} (w - 5) = -10$.

> The graph is just the line $y = w - 5$ with the point $(-5, -10)$ deleted.

5. $\log_8 1 = 0$.

6. $\dfrac{2}{0.5} = 4$.

7. The limit does not exist.

> Contrast problem #6 and problem #7.
> The function $f(x) = \dfrac{2}{x - 2}$ is defined at $x = 2.5$, but not at $x = 2$.

8. $\lim\limits_{x \to 2} \dfrac{(x - 2)(x - 1)}{(x - 2)(x + 1)} = \lim\limits_{x \to 2} \dfrac{x - 1}{x + 1} = \dfrac{1}{3}$.

9. $\lim\limits_{h \to 0} \dfrac{2xh + h^2}{h} = \lim\limits_{h \to 0} (2x + h) = 2x$.

10. $\lim\limits_{x \to 1} \dfrac{x^2(x - 1)}{x - 1} = \lim\limits_{x \to 1} x^2 = 1$.

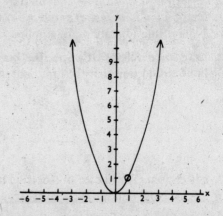

> This function with its complicated-looking equation is just the parabola $y = x^2$ with the point $(1, 1)$ deleted.

11. $\lim\limits_{x \to 0} f(x)$ does not exist.

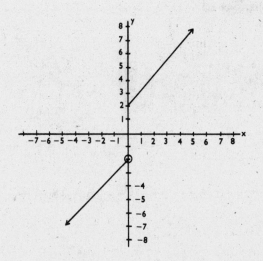

As x approaches 0 from the right, $\lim_{x \to 0^+} f(x) = 2$. As x approaches 0 from the left, $\lim_{x \to 0^-} f(x) = -2$. There is no unique L such that $\lim_{x \to 0} f(x) = L$; hence, the limit as x approaches 0 does not exist.

*Section 2.1A: Epsilon-Delta Definition of Limit

Definition (precise): The statement

$$\lim_{x \to a} f(x) = L$$

means that corresponding to a positive number ε, arbitrarly chosen, there can be found a positive number δ, such that if

$$0 < |x - a| < \delta,$$

then

$$|f(x) - L| < \varepsilon.$$

Graphic illustration of epsilon-delta definition:

Figure 1. $\lim_{x \to a} f(x) = L$. In this example, $f(a)$ is not defined. It is *extremely important* to realize that a function does not have to be defined at a specific point in order to have a limit at that point.

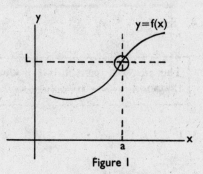

Figure 1

Figure 2. ε is chosen. The choice of ε determines b_1 and b_2 on the x-axis. Note that if $b_1 < x < b_2$, then $|f(x) - L| < \varepsilon$.

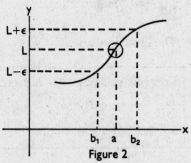

Figure 2

Figure 3. Choose $\delta > 0$ to be less than the smaller of $|b_1 - a|$, $|b_2 - a|$. This will determine an open interval $(a - \delta, a + \delta)$ such that $|f(x) - L| < \varepsilon$ whenever $0 < |x - a| < \delta$.

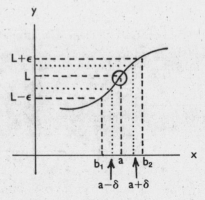

Figure 3

Q1. Prove, using the precise definition of limit, that $\lim_{x \to 5} 2x = 10$.

> Answer: One need only show that for $\varepsilon > 0$, there exists $\delta > 0$ such that if $0 < |x - 5| < \delta$, then $|2x - 10| < \varepsilon$. Note that $|2x - 10| = 2|x - 5| < \varepsilon$ if $|x - 5| < \varepsilon/2$. Choose $\delta = \varepsilon/2$.

Authors' Note: In Q1, we could choose any δ satisfying $0 < \delta < \dfrac{\varepsilon}{2}$. It often helps to think of ε, δ proofs as a challenge. That is, someone provides you with a small positive ε and you must then produce a small positive δ (which usually will depend on ε).

Q2. Given $\varepsilon = 0.006$, find a δ to establish $\lim_{x \to 3} \dfrac{3x^2 - 27}{x - 3} = 18$. _____

> Answer: If $x \neq 3$, then $\left| \dfrac{3x^2 - 27}{x - 3} - 18 \right| = \left| \dfrac{3(x + 3)(x - 3)}{x - 3} - 18 \right| = |3x + 9 - 18| =$
> $|3x - 9| = 3|x - 3| < 0.006$ if $|x - 3| < 0.002$.
> Hence $\delta = 0.001$ would suffice (as would any positive number less than 0.002).

Problem Set 2.1A

1. For each limit statement, find a value of δ corresponding to $\varepsilon = 0.01$.
   (a) $\lim_{x \to 6} 7x = 42$

   (b) $\lim_{x \to 2} \dfrac{x^2 - 4}{x - 2} = 4$

2. Using the precise definition of limit, prove
   (a) $\lim_{x \to 2} (3x - 5) = 1$

   (b) $\lim_{x \to 0} \dfrac{x^2 - x}{x} = -1$

   (c) $\lim_{x \to 3} \dfrac{1}{x} = \dfrac{1}{3}$

   (d) $\lim_{x \to 2} x^3 = 8$

3. In each case, find the indicated limit, if it exists. If no limit exists, sketch a graph of the function

to demonstrate the lack of a limit at the specific point.

(a) $\lim\limits_{x \to \pi/4} f(x)$, where $f(x) = \sin x + \cos x$

(b) $\lim\limits_{x \to 0} f(x)$, where $f(x) = \begin{cases} x^2 & \text{if } x \geq 0 \\ -1 & \text{if } x < 0 \end{cases}$

(c) $\lim\limits_{x \to 1} f(x)$, where $f(x) = \begin{cases} x & \text{if } x \geq 1 \\ 2 - x & \text{if } x < 1 \end{cases}$

(d) $\lim\limits_{x \to 0} f(x)$, where $f(x) = [\![x]\!]$

(greatest integer function)

Solutions and Comments for Problem Set 2.1A

1. (a) $|7x - 42| = 7|x - 6| < 0.01$ if
$|x - 6| < \frac{1}{7}(0.01) = 0.0014285$
(approximately).
Choose $\delta = 0.001$.

(b) $|(x + 2) - 4| = |x - 2| < 0.01$.
Choose $\delta = 0.005$.

> There are many possible choices for δ in (a) and (b). Also, note that $f(x) = 7x$ is defined at $x = 6$ and that $f(x) = \dfrac{x^2 - 4}{x - 2}$ is not defined at $x = 2$.

2. (a) $|(3x - 5) - 1| < \varepsilon$ if
$|3x - 6| = 3|x - 2| < \varepsilon \Rightarrow |x - 2| < \varepsilon/3$.
Choose $\delta = \varepsilon/3$.

(b) $\left| \dfrac{x^2 - x}{x} - (-1) \right| < \varepsilon$ if $|(x - 1) + 1| =$
$|x - 0| < \varepsilon$. Choose $\delta = \varepsilon$.

(c) $\left| \dfrac{1}{x} - \dfrac{1}{3} \right| = \left| \dfrac{3 - x}{3x} \right| = \left| \dfrac{x - 3}{3x} \right|$. If x is
restricted to the interval $(2, 4)$, then
$\left| \dfrac{x - 3}{12} \right| < \left| \dfrac{x - 3}{3x} \right| < \left| \dfrac{x - 3}{6} \right|$. Hence
$\left| \dfrac{x - 3}{3x} \right| < \varepsilon$ if $\left| \dfrac{x - 3}{6} \right| < \varepsilon \Rightarrow |x - 3| < 6\varepsilon$.

Choose δ to be the smaller of $1, 6\varepsilon$.

> Students sometimes get confused when confronted with limit problems like those in (c) and (d). If you are attempting to find an interval containing 3, as is the case in (c), you can certainly require that x be in the open interval $(2, 4)$ unless, for some reason, the interval containing 3 must be at least 2 units in length.

(d) $x^3 - 8 = (x - 2)(x^2 + 2x + 4)$. Hence
$|x^3 - 8| < \varepsilon$ if $|(x - 2)(x^2 + 2x + 4)| < \varepsilon$. If
x is restricted to the interval $(1, 3)$, then
$|(x - 2) \times 7| < |(x - 2)(x^2 + 2x + 4)| <$
$|(x - 2) \times 19|$. Hence $|x^3 - 8| < \varepsilon$ if
$|19(x - 2)| < \varepsilon \Rightarrow |x - 2| < \varepsilon/19$. Choose δ
to be the smaller of $1, \varepsilon/19$.

3. (a) $\dfrac{\sqrt{2}}{2} + \dfrac{\sqrt{2}}{2} = \sqrt{2}$;

(b) No limit;

(c) 1;

(d) No limit.

> Functions (a) and (c) are continuous throughout their domains. The function in (b) has one point of discontinuity (at $x = 0$). The function in (d) is discontinuous at every integer value.

Section 2.2: The Number *e*

The number *e*, like π, is irrational ... and extremely important in mathematics.

$$e = \lim_{n \to \infty} \left(1 + \frac{1}{n}\right)^n$$

Authors' Note: A simple BASIC computer program that demonstrates the convergence of $\left(1 + \frac{1}{n}\right)^n$ as *n* increases is shown on the right.

```
10  FOR N = 1 TO 1000
20  E = (1 + 1/N) ∧ N
30  PRINT E
40  NEXT N
50  END
```

The value of *e*, to eight significant figures, is 2.7182818. The number *e* can also be defined as the sum of an infinite series:

$$e = \sum_{k=0}^{\infty} \frac{1}{k!} = 1 + \frac{1}{1!} + \frac{1}{2!} + \frac{1}{3!} + \frac{1}{4!} + \cdots$$

Also,

$$e^x = \sum_{k=0}^{\infty} \frac{1}{k!} x^k = 1 + \frac{x}{1!} + \frac{x^2}{2!} + \frac{x^3}{3!} + \frac{x^4}{4!} + \cdots$$

The number *e* is the base for *natural* logarithms: $\log_e x = \ln x$. Note that $\ln e = 1$.

Q1. Calculate: (a) $\ln e^3$ _____   (b) $\ln \sqrt[5]{e}$ _____

(c) $\dfrac{\ln 1000}{\ln 10}$ _____

Answers: (a) $3 \times \ln e = 3 \times 1 = 3$.   (b) $\frac{1}{5} \ln e = \frac{1}{5}$.   (c) $\log_{10} 1000 = 3$.

Authors' Note: Remember that $\dfrac{\log_b a}{\log_b c} = \log_c a$. We could calculate $\ln 26$ from a table of common logarithms by writing $\ln 26 = \dfrac{\log_{10} 26}{\log_{10} e}$.

Problem Set 2.2

1. On the same set of axes, sketch the graphs of $y = e^x$ and $y = \ln x$.

2. Calculate:
   (a) $\ln e^{32}$

(b) $e^{\ln 3}$
(c) $\ln \sqrt{e^5}$

3. Sketch the graph of $y = e^{-x}$.

4. Express e^{-x} as the sum of an infinite series.

Solutions and Comments for Problem Set 2.2

1.

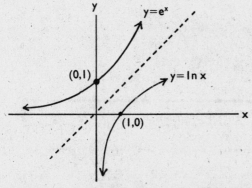

The two functions are inverses; hence, their graphs are symmetric to the line $y = x$.

3.

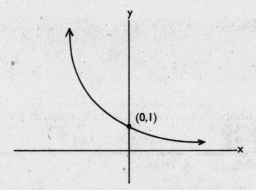

Common error: $e^{-x} = -e^x$.

2. (a) $32 \ln e = 32$;
   (b) 3;
   (c) $\frac{1}{2} \times \ln e^5 = \frac{1}{2} \times 5 = 2\frac{1}{2}$.

(b) is just "e raised to the power to which e must be raised in order to obtain 3."

4. $\sum_{k=0}^{\infty} \frac{(-1)^k}{k!} x^k = 1 - \frac{x}{1!} + \frac{x^2}{2!} - \frac{x^3}{3!} + \cdots$

In the infinite series representation for e^x, substitute $-x$ for x.

Section 2.3: A Very Important Limit

Consider the central angle θ (in radians) in the circle on the right.

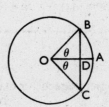

Since θ is in radians, then the length of arc $\overset{\frown}{BA}$ is $r\theta$ units; hence $\overset{\frown}{BC} = 2r\theta$. Also, $BD = r\sin\theta$ and thus $BC = 2r\sin\theta$.

Hence, $\dfrac{\sin\theta}{\theta} = \dfrac{2r\sin\theta}{2r\theta} = \dfrac{BC}{\text{arc } \overset{\frown}{BC}}$. Now, the limit of the ratio of a chord of a circle to its arc, as the length

of the arc approaches zero, is 1. Hence $\lim_{\theta\to 0}\dfrac{\sin\theta}{\theta} = \lim_{\theta\to 0}\dfrac{BC}{\text{arc } \overset{\frown}{BC}} = 1$. The important conclusion:

$$\lim_{\theta\to 0}\frac{\sin\theta}{\theta} = 1$$

Authors' Note: If degree measure is used, the limit is $\pi/180$.

Q1. Find: (a) $\lim\limits_{\theta \to 0} \dfrac{\sin \theta}{\theta \sec \theta}$ _____   (b) $\lim\limits_{\beta \to 0} \dfrac{1 - \cos^2 \beta}{\beta}$ _____

Answers: (a) $\lim\limits_{\theta \to 0} \dfrac{\sin \theta}{\theta} \times \cos \theta = 1 \times 1 = 1.$ (b) $\lim\limits_{\beta \to 0} \dfrac{\sin^2 \beta}{\beta} = \lim\limits_{\beta \to 0} \dfrac{\sin \beta}{\beta} \times \sin \beta = 1 \times 0 = 0.$

Problem Set 2.3

Find the indicated limits.

1. $\lim\limits_{x \to \pi} \dfrac{\sin x}{x}$

2. $\lim\limits_{\theta \to \pi} \dfrac{\cos \theta}{\theta}$

3. $\lim\limits_{\theta \to 0} \dfrac{\sin 2\theta}{\theta}$

4. $\lim\limits_{\theta \to 0} \dfrac{\tan \theta \cos \theta}{\theta}$

Solutions and Comments for Problem Set 2.3

1. $\dfrac{0}{\pi} = 0.$

Note that the function $f(x) = \dfrac{\sin x}{x}$ is defined for all numbers except $x = 0$.

2. $\dfrac{-1}{\pi}.$

3. $\lim\limits_{\theta \to 0} \dfrac{2 \sin \theta \cos \theta}{\theta} = \lim\limits_{\theta \to 0} 2 \left(\dfrac{\sin \theta}{\theta} \right) \cos \theta$

$$= 2 \times 1 \times 1 = 2.$$

4. $\lim\limits_{\theta \to 0} \left(\dfrac{\sin \theta}{\cos \theta} \times \dfrac{\cos \theta}{\theta} \right) = \lim\limits_{\theta \to 0} \dfrac{\sin \theta}{\theta} = 1.$

Section 2.4: Non-Existent Limits

Given a function f, $\lim\limits_{x \to a} f(x)$ may not exist. That is, $f(x)$ may not approach a unique value, L, as x approaches a.

Q1. Find: (a) $\lim\limits_{x \to 8} \dfrac{1}{x^2}$ _____   (b) $\lim\limits_{x \to 0} \dfrac{1}{x^2}$ _____

Answers: (a) $\dfrac{1}{8^2} = \dfrac{1}{64}.$ (b) Limit does not exist.

Authors' Note: The graph of $y = \dfrac{1}{x^2}$ is on the right. Note that $\dfrac{1}{x^2}$ becomes infinite as $x \to 0$. The function does not approach a unique number L as $x \to 0$.

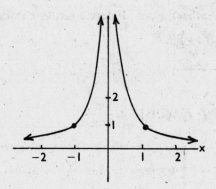

Q2. Why is $\lim\limits_{x \to 0} \sin \dfrac{1}{x}$ non-existent? _____

Answer: Any interval containing 0 will find $\sin \dfrac{1}{x}$ ranging over all values in the closed interval $[-1, 1]$.

Authors' Note: Referring to Q2, consider the interval $(-1/100, 1/100)$. The numbers $a = \dfrac{1}{100\pi}$, $b = \dfrac{1}{100\pi + \frac{\pi}{2}}$, and $c = \dfrac{1}{100\pi + \frac{3\pi}{2}}$ are all in the given interval about zero. Note that $\sin \dfrac{1}{a} = 0$, $\sin \dfrac{1}{b} = 1$, and $\sin \dfrac{1}{c} = -1$.

Q3. Why is $\lim\limits_{x \to 0} \dfrac{|x|}{x}$ non-existent? _____

Answer: The graph is shown on the right. Note that $\dfrac{|x|}{x}$ does not approach a unique value as $x \to 0$.

Problem Set 2.4

In problems 1–7, find all real values of c for which $\lim\limits_{x \to c} f(x)$ is non-existent.

1. $f(x) = \dfrac{1}{x}$

2. $f(x) = \dfrac{5}{x - 4}$

3. $f(x) = \dfrac{x - 2}{(x + 4)(x - 7)}$

4. $f(x) = |x|$

5. $f(x) = \tan x$

6. $f(x) = \dfrac{\cos x}{x}$

7. $f(x) = \begin{cases} x^2 & \text{if } x \geq 2 \\ -1 & \text{if } x < 2 \end{cases}$

In problems 8–10, sketch the graph of the function to demonstrate that the indicated limit does not exist.

8. $\lim\limits_{x \to 0} \dfrac{1}{x^3}$

9. $\lim\limits_{x \to 2} [\![x]\!]$ (greatest integer function)

10. $\lim\limits_{x \to 3} \dfrac{|x - 3|}{x - 3}$

Solutions and Comments for Problem Set 2.4

1. 0.

2. 4.

3. −4, 7.

Common error:
$$\lim\limits_{x \to 2} \frac{x - 2}{(x + 4)(x - 7)} \text{ does not exist.}$$
Note that $f(2) = 0$ and that
$$\lim\limits_{x \to 2} \frac{x - 2}{(x + 4)(x - 7)} = 0.$$

4. There are no values for which $\lim\limits_{x \to c} |x|$ does not exist.

5. Limit does not exist for all numbers of the form $(2n + 1)\frac{\pi}{2}$, where n is an integer.

6. 0.

7. 2.

8.

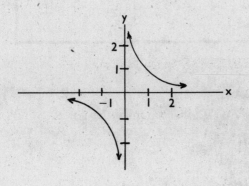

9.

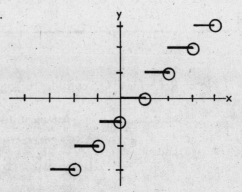

Note that if c is any non-integer, then $\lim\limits_{x \to c} [\![x]\!]$ exists.

10.

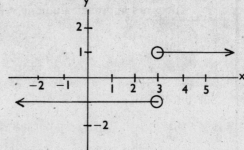

Section 2.5: Continuity

A function *f* is said to be *continuous* at $x = c$ if
   (a) $f(c)$ is defined;
   (b) $\lim\limits_{x \to c} f(x)$ exists;

   (c) $\lim\limits_{x \to c} f(x) = f(c)$.

In any of these conditions fails to be satisfied, the function is *discontinuous* at $x = c$.

The following illustrations demonstrate some of the varieties of discontinuities that can exist.

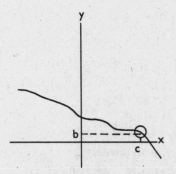

$\lim\limits_{x \to c} f(x) = b$, but $f(c)$ is not defined.

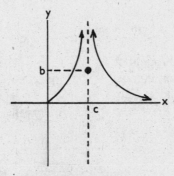

$f(c) = b$, but $\lim\limits_{x \to c} f(x)$ does not exist.

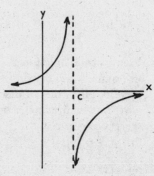

$\lim\limits_{x \to c} f(x)$ does not exist and $f(c)$ is not defined.

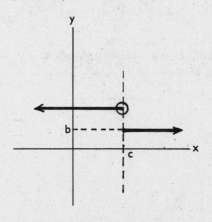

$f(c) = b$, but $\lim\limits_{x \to c} f(x)$ does not exist.

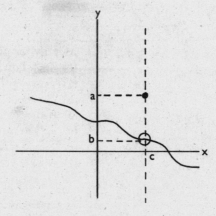

$f(c) = a$ and $\lim\limits_{x \to c} f(x) = b$, but $\lim\limits_{x \to c} f(x) \neq f(c)$.

A function that is continuous at every point in its domain is called a *continuous function*.

Q1. If $f(x) = \begin{cases} \dfrac{x^2 - 36}{x - 6} & \text{if } x \neq 6 \\ k & \text{if } x = 6 \end{cases}$, what must be the value of *k* if *f* is to be continuous at $x = 6$?

$$\text{Answer: } \lim_{x \to 6} \frac{x^2 - 36}{x - 6} = \lim_{x \to 6} \frac{(x + 6)(x - 6)}{x - 6} = \lim_{x \to 6} (x + 6) = 12; \text{ hence } k = 12.$$

Authors' Notes: The graph of $f(x) = \dfrac{x^2 - 36}{x - 6}$ is just the line $y = x + 6$ with the point $(6, 12)$ deleted. If we define $f(6)$ to be 12, we "fill the hole" at $(6, 12)$.

A rule of thumb: If one can draw the graph of a function without lifting a pen or pencil from the paper, then the function is continuous.

Problem Set 2.5

1. Identify each function as continuous throughout its domain or as discontinuous. If discontinuous, identify all points of discontinuity and explain why the function is discontinuous.

   (a) $f(x) = \sin x$

   (b) $g(x) = |x|$

   (c) $f(x) = \dfrac{|x|}{x}$

   (d) $h(x) = \ln x$

   (e) $w(x) = \dfrac{(x - 3)(x + 1)}{x + 1}$

   (f) $f(x) = \begin{cases} \dfrac{x^2 - 4}{x - 2} & \text{if } x \neq 2 \\ 2 & \text{if } x = 2 \end{cases}$

   (g) $f(x) = \begin{cases} x^2 - 1 & \text{if } x \geq 0 \\ x - 1 & \text{if } x < 0 \end{cases}$

2. Find the value of k, if it exists, that will make each function continuous at $x = 1$.

   (a) $f(x) = \begin{cases} \dfrac{x^4 - 1}{x - 1} & \text{if } x \neq 1 \\ k & \text{if } x = 1 \end{cases}$

   (b) On the open interval $(0, 2)$
   $$f(x) = \begin{cases} [\![x]\!] & \text{if } x \text{ is not an integer} \\ k & \text{if } x \text{ is an integer} \end{cases}$$
   $[\![x]\!]$ is the greatest integer function.

   (c) $g(x) = \begin{cases} x^2 + 7 & \text{if } x \geq 1 \\ x + k & \text{if } x < 1 \end{cases}$

3. Identify the points of discontinuity, if they exist, for the following trigonometric functions.

   (a) $f(x) = \sin x$

   (b) $g(x) = \cos x$

   (c) $t(x) = \tan x$

   (d) $s(x) = \sec x$

   (e) $c(x) = \csc x$

   (f) $w(x) = \cot x$

Solutions and Comments for Problem Set 2.5

1. (a) continuous;

   (b) continuous;

   (c) discontinuous at $x = 0$, $f(0)$ not defined;

   (d) continuous;

   (e) discontinuous at $x = -1$, $f(-1)$ not defined;

   (f) discontinuous at $x = 2$, $f(2) = 2$ and $\lim_{x \to 2} f(x) = 4$;

   (g) continuous.

In (f), the graph of $f(x) = \dfrac{x^2 - 4}{x - 2}$ is the line $y = x + 2$ with the point $(2, 4)$ deleted. Defining $f(2)$ to be 2 doesn't fill the hole. In (g), $f(0) = -1$ and $\lim_{x \to 0} g(x) = -1$. The two portions of the graph "hook up" at $(0, -1)$.

2. (a) $\lim_{x \to 1} \dfrac{(x^2 + 1)(x + 1)(x - 1)}{x - 1}$

$= \lim_{x \to 1}(x^2 + 1)(x + 1) = 2 \times 2 = 4.$

(b) No value of k will make the function continuous. The graph of f is shown below.

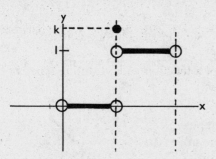

(c) The two portions of the graph must "hook up" at $(1, 8)$. Hence $1 + k = 8 \Rightarrow k = 7$.

3. (a) none;

(b) none;

(c) discontinuous at $x = (2n + 1)\frac{\pi}{2}$, where n is any integer;

(d) same as (c);

(e) discontinuous at $x = n\pi$, where n is any integer;

(f) same as (e).

> Since $\cot x = \dfrac{1}{\tan x}$, $y = \cot x$ is discontinuous when $\tan x = 0$. Since $\tan x = \dfrac{\sin x}{\cos x}$, $y = \tan x$ is discontinuous when $\cos x = 0$. Knowledge of the graphs of the trigonometric functions helps to identify discontinuities.

Section 2.6: Important Continuity Theorems

Theorem I (*Intermediate Value Theorem*): If f is continuous on the closed interval $[a, b]$ and if $f(a) = A$ and $f(b) = B$, then corresponding to any $C \in [A, B]$, there exists at least one $c \in [a, b]$ such that $f(c) = C$.

Q1. Given $f(x) = x^{3/2} + x^{1/2} - 3$. Prove that there exists $c \in [1, 4]$ such that $f(c) = 0$. _____

> Answer: f is continuous on $[1, 4]$, $f(1) = -1$, $f(4) = 4^{3/2} + 4^{1/2} - 3 = 8 + 2 - 3 = 7$. Since $0 \in [-1, 7]$, the Intermediate Value Theorem states that there exists at least one $c \in [1, 4]$ such that $f(c) = 0$.

Authors' Note: A continuous function has the property that one can draw its complete graph without lifting pencil or pen from the paper. In Q1, to draw the graph from $(1, -1)$ to $(4, 7)$ without lifting the writing implement, one would have to cross (intersect) the x-axis at least once. Hence, there would be at least one c such that $y = 0 = f(c)$.

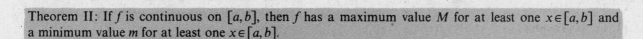

Theorem II: If f is continuous on $[a, b]$, then f has a maximum value M for at least one $x \in [a, b]$ and a minimum value m for at least one $x \in [a, b]$.

Q2. Which of the following functions have a maximum and a minimum value in $[0, 2]$?

(a) $f(x) = x^2 \sin x$

(b) $f(x) = 3x$

(c) $f(x) = \dfrac{1}{(x - 1)^2}$

(d) $g(x) = \begin{cases} x^2 + 4 & \text{if } x \geq 1 \\ 5 & \text{if } x < 1 \end{cases}$ _____

Answers: (a), (b), and (d) are continuous on $[0, 2]$ and hence have a maximum and minimum value on that interval by Theorem II. The function in (c) is not continuous at $x = 1$ and $\lim\limits_{x \to 1} \dfrac{1}{(x - 1)^2} = \infty$.

 Common error: Consider $t(x) = \begin{cases} x & \text{if } x \geq 1 \\ -x & \text{if } x < 1 \end{cases}$. Since t is discontinuous at $x = 1$, it is not continuous over the interval $[0, 2]$ and hence has no maximum or minimum value on that interval.

Authors' Note: It is true that t is not continuous over $[0, 2]$, but it does have a maximum value of 2 and a minimum value of -1 (see graph.) Theorem II does not say that only continuous functions have maximum and minimum values over a closed interval.

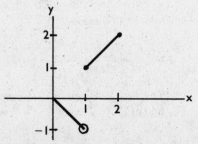

Problem Set 2.6

1. Given $f(x) = \dfrac{x^3 + x - 12}{x - 1}$. Prove that there is at last one $c \in [2, 3]$ such that $f(c) = 0$.

2. If $g(x) = x^2 - 3x + 2$, prove that $g(x)$ assumes all values in the interval $[2, 12]$ as x assumes all values in $[3, 5]$.

3. Which of the following functions are guaranteed a maximum and minimum value on the specified interval by Theorem II?
   (a) $f(x) = x^2 - 6$ on $[-3, 2]$
   (b) $g(x) = (\sin x)(\cos x)$ on $[0, \pi]$
   (c) $h(x) = \dfrac{|x - 1|}{x - 1}$ on $[-3, 6]$
   (d) $t(x) = \tan x$ on $[0, \pi]$
   (e) $z(x) = 6$ on $[-1000, 1000]$
   (f) $r(x) = \begin{cases} \sin x & \text{if } x \geq 0 \\ -x & \text{if } x < 0 \end{cases}$ on $\left[-\dfrac{\pi}{2}, \dfrac{\pi}{2}\right]$

Solutions and Comments for Problem Set 2.6

1. f is continuous on $[2, 3]$, $f(2) = -2$, $f(3) = 9$. Since $0 \in [-2, 9]$, the Intermediate Value Theorem guarantees a $c \in [2, 3]$ such that $f(c) = 0$.

> The Intermediate Value Theorem could not be applied to any interval containing 1 since f is not continuous at $x = 1$.

2. g is continuous on $[3, 5]$, $g(3) = 2$, $g(5) = 12$. For any $y_1 \in [2, 12]$, the Intermediate Value Theorem guarantees that for at least one $x_1 \in [3, 5]$, $g(x_1) = y_1$.

> g is continuous over the set of real numbers.

3. (a), (b), (e), and (f).

> In (c), h has a maximum value of 1 and a minimum value of -1, but Theorem II does not apply since h is discontinuous at $x = 1$. In (d), the function is discontinuous at $x = \pi/2$. In (e), the maximum and minimum values are both 6. In (f), note that $r(0) = 0$ and $\lim\limits_{x \to 0} r(x) = 0$; hence r is continuous at 0.

Section 2.7: Multiple Choice Questions for Chapter 2

1. $\lim\limits_{t\to 0}\dfrac{\sin t}{t} =$
   (A) 0
   (B) 1
   (C) $\pi/2$
   (D) -1
   (E) undefined

2. $\ln_e 10 =$
   (A) $\ln_{10} e$
   (B) $\dfrac{1}{\log_{10} e}$
   (C) e^{10}
   (D) $\sqrt[10]{e}$
   (E) $10(\ln e)$

3. If f is continuous on $[-4, 4]$ such that $f(-4) = 11$ and $f(4) = -11$, then
   (A) $f(0) = 0$
   (B) $\lim\limits_{x\to 2} f(x) = 8$
   (C) There is at least one $c \in [-4, 4]$ such that $f(c) = 8$
   (D) $\lim\limits_{x\to 3} f(x) = \lim\limits_{x\to -3} f(x)$
   (E) It is possible that f is not defined at $x = 0$.

4. $\lim\limits_{\theta\to 0}\dfrac{\sin\theta}{\sec\theta} =$
   (A) 1
   (B) 0
   (C) $\pi/2$
   (D) -1
   (E) ∞

5. If $f(x) = \begin{cases} \dfrac{x^2 - 4}{x - 2} & \text{if } x \neq 2 \\ k & \text{if } x = 2 \end{cases}$,

   for what value(s) of k is f continuous at $x = 2$?
   (A) $-2, 2$
   (B) 4
   (C) 8
   (D) 0
   (E) 6

6. Which graph best represents the function $y = \dfrac{x^3 - 3x^2}{x^2}$?

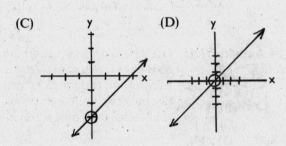

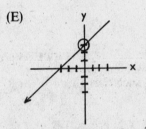

7. If $\lim\limits_{x\to c} f(x) = 0$ and $\lim\limits_{x\to c} g(x) = \infty$, then
   $\lim\limits_{x\to c} \big[f(x)g(x) \big] =$
   (A) 0
   (B) 1
   (C) ∞
   (D) $-\infty$
   (E) Not enough information to evaluate the indicated limit.

8. $\lim\limits_{x\to 5}\dfrac{x - 5}{x - 5} =$
   (A) 0
   (B) 1
   (C) -1
   (D) $\frac{1}{2}$
   (E) undefined

9. $\ln e^4 =$
   (A) 4
   (B) 10^4
   (C) $4(\ln 10)$
   (D) e^4
   (E) $4e$

10. If f is continuous on $[4,7]$, *how many* of the following statements must be true?
    (i) f has a maximum value on $[4,7]$.
    (ii) f has a minimum value on $[4,7]$.
    (iii) $f(7) > f(4)$.
    (iv) $\lim_{x \to 6} f(x) = f(6)$.

    (A) 0
    (B) 1
    (C) 2
    (D) 3
    (E) 4

11. $\lim_{x \to \pi/4} (\sin^2 x - 1) =$

    (A) 0
    (B) 1
    (C) -1
    (D) $\pi/4$
    (E) $-\frac{1}{2}$

12. Referring to the following figure showing the graph of $y = f(x)$, $\lim_{x \to c} f(x) =$

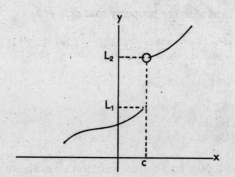

    (A) L_1
    (B) L_2
    (C) $\dfrac{L_1 + L_2}{2}$
    (D) $L_1 + L_2$
    (E) undefined

13. $\lim_{x \to 5} f(x)$ is non-existent if $f(x) =$

    (A) $\dfrac{x - 5}{5}$
    (B) $\sin (x - 5)$
    (C) $\ln x$
    (D) $\dfrac{5x}{x + 5}$
    (E) $\dfrac{x}{x^2 - 5x}$

14. If $f(x) = \begin{cases} 2x^2 + 3 & \text{if } x \geq 1 \\ g(x) & \text{if } x < 1 \end{cases}$,
    then f will be continuous at $x = 1$ if $g(x) =$
    (A) x
    (B) $\cos (x + 4)$
    (C) $6 - x$
    (D) $x^2 + 2$
    (E) $2x^2 - 3$

*15. In proving that $\lim_{x \to 7} 3x = 21$, what is the largest possible choice for δ if one chooses an $\varepsilon > 0$ such that $|3x - 21| < \varepsilon$.
    (A) ε
    (B) $\varepsilon/2$
    (C) $\varepsilon/3$
    (D) $\varepsilon/7$
    (E) $\varepsilon/21$

Solutions and Comments for Multiple Choice Questions

1. **(B)**

> Famous limit ... see Section 2.3.

2. **(B)**

> $\log_b a = \dfrac{1}{\log_a b}$, a basic law of logarithms.

3. **(C)** Apply the Intermediate Value Theorem.

4. **(B)** $\lim_{\theta \to 0} (\sin \theta)(\cos \theta) = 0 \times 1 = 0$.

> Since $\lim_{\theta \to 0} \sin \theta$ and $\lim_{\theta \to 0} \cos \theta$ both exist, rule (3) of Section 2.1 can be applied.

5. **(B)** $\lim_{x \to 2} \dfrac{(x + 2)(x - 2)}{x - 2} = \lim_{x \to 2} (x + 2) = 4$.

6. **(C)** $y = \dfrac{x^2(x - 3)}{x^2} = x - 3$ if $x \neq 0$.

> Graph is the line $y = x - 3$ with the point $(0, -3)$ deleted.

7. **(E)** Can't apply rule (3) of Section 2.1 since $\lim_{x \to c} g(x)$ does not exist.

> Review Q3 in Section 2.1 if you missed this problem.

8. **(B)** The graph of $y = \dfrac{x - 5}{x - 5}$ is just the horizontal line $y = 1$ with the point $(5, 1)$ deleted.

9. **(A)** $\ln e^4 = 4(\ln e) = 4 \times 1 = 4$.

> $\log_b a^c = c \times \log_b a$.

10. **(D)** Statement (iii) is not necessarily true.

> Statements (i) and (ii) are a result of Theorem II in Section 2.6. Statement (iv) is true because of the definition of continuity.

11. **(E)** Given limit is
$$\left(\frac{1}{\sqrt{2}}\right)^2 - 1 = \frac{1}{2} - 1 = -\frac{1}{2}.$$

> $y = \sin^2 x - 1$ is continuous at $x = \pi/4$.

12. **(E)** Review Section 2.5 if you had difficulty with this problem.

13. **(E)**

> **(E)** is the only answer choice with a zero denominator if $x = 5$.

14. **(C)** Since $\lim_{x \to 1} (2x^2 + 3) = 5$, g must be a function with the property that $g(1) = 5$.

15. **(C)** $|3x - 21| = 3|x - 7| < \varepsilon$ if $|x - 7| < \varepsilon/3$.

Chapter 3
DERIVATIVES

Section 3.1: Definition of the Derivative

If f is a function that is continuous on an interval $[x, x + h]$, then the *average rate of change* of f on the interval is

$$\frac{f(x + h) - f(x)}{h}.$$

Q1. Given $f(x) = 2x - x^2$, what is the average rate of change of f on $[-1, 4]$? _____

Answer: $\dfrac{f(4) - f(-1)}{5} = \dfrac{-8 - (-3)}{5} = -1.$

Authors' Note: The average rate of change on any interval $[a, b]$ is just the slope of the line joining the points $(a, f(a))$ and $(b, f(b))$.

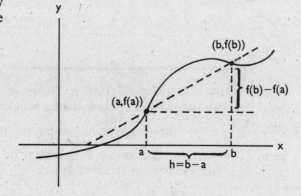

The *instantaneous rate of change* of f at a point in its domain is called the *derivative* of f and is defined to be

$$f'(x) = \lim_{h \to 0} \frac{f(x + h) - f(x)}{h} \quad \text{if this limit exists.}$$

Authors' Note: The derivative is a derived function of f which gives the slope of the tangent line to the curve if the tangent line is not vertical. As the figure at the top of page 67 indicates, the tangent line at A is the limit of the secant line connecting points A and B as B approaches A, that is, as $h \to 0$.

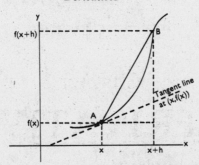

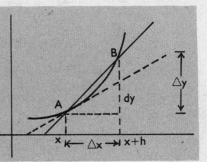

The derivative can be written several ways. Some common notations include

$$f'(x) = \frac{dy}{dx} = y' = \frac{d}{dx}(f(x)) = D_x y$$

The derivative represents the rate of change of y with respect to a point in the domain. If $\Delta y = f(x + h) - f(x)$ and $\Delta x = h$, then

$$\frac{dy}{dx} = \lim_{\Delta x \to 0} \frac{\Delta y}{\Delta x}.$$

Q2. (a) Given that $f(x) = 3x^2 - 2x - 1$, calculate $f(x + h)$. _____

(b) Calculate $f'(x)$. _____

(c) Find the slope of the tangent line to f at the point where $x = -1$. _____

Answer: (a) $3(x + h)^2 - 2(x + h) - 1 = 3x^2 + 6xh + 3h^2 - 2x - 2h - 1$.

(b) $\lim_{h \to 0} \dfrac{f(x + h) - f(x)}{h} = \lim_{h \to 0} \dfrac{6xh + 3h^2 - 2h}{h} = 6x - 2 = f'(x)$.

(c) $f'(-1) = -8$.

The derivative of the function $f(x) = ax^n$ is very important in calculus. It is listed in the next section and we will give a justification for the formula here. Using the binomial expansion formula, $(x + h)^n = x^n + nx^{n-1}h + \dfrac{n(n - 1)x^{n-2}h^2}{2!} + \cdots + h^n$. Without getting too formal, let's just note that from the third term on, each term has at least two factors of h so we will write $(x + h)^n = x^n + nx^{n-1}h + h^2 w$, where w is "what is left" when h^2 is factored from these terms. Hence

$$f'(x) = \lim_{h \to 0} \frac{a(x + h)^n - ax^n}{h} = \lim_{h \to 0} \frac{a[(x + h)^n - x^n]}{h}$$

$$= \lim_{h \to 0} \frac{a[x^n + nx^{n-1}h + h^2 w - x^n]}{h} = \lim_{h \to 0} \frac{a[nx^{n-1}h + h^2 w]}{h}$$

$$= \lim_{h \to 0} a[nx^{n-1} + hw] = anx^{n-1} + a \times 0 \times w = nax^{n-1} + 0 = nax^{n-1}$$

Q3. If $f(x) = 4x^9$, find $f'(x)$. _____

Answer: Using the formula above, $f'(x) = 9 \times 4x^{9-1} = 36x^8$.

The formula is valid for any real number n. Hence, if $f(x) = x^{1/4}$, then $f'(x) = \frac{1}{4}x^{-3/4}$.

Problem Set 3.1

1. If $f(x) = \sin x + \cos x$, what is the average rate of change of $f(x)$ with respect to x on the interval $\left[-\frac{\pi}{4}, \frac{\pi}{4}\right]$?

2. A particle moves in a straight line according to the distance formula $s(t) = t^3 - 2t^2$, where s is measured in feet and t is time measured in seconds. What is the average velocity of the particle between $t = 3$ and $t = 5$?

3. Given $f(x) = 2x^2 + 1$.
   (a) Find $f'(x)$ using the definition of derivative.
   (b) What is the slope of the tangent line to the curve at $x = -2$?

(c) What is the instantaneous rate of change of f with respect to x when $x = 0$?

4. Use the definition of the derivative to calculate $f'(1)$ and $f'(0)$ if $f(x) = \frac{1}{x}$.

5. Use the definition of derivative to calculate $f'(x)$ if $f(x) = \sqrt{x}$.

6. Use the formula developed in the reading section to calculate the derivative of
   (a) $-3x^2$
   (b) $7x^{-2}$
   (c) $9x^{1/3}$

Solution and Comments for Problem Set 3.1

1. $\dfrac{f(\pi/4) - f(-\pi/4)}{\pi/2} = \dfrac{2\sqrt{2}}{\pi}$.

$$\sin\frac{\pi}{4} = \cos\frac{\pi}{4} = \cos\left(-\frac{\pi}{4}\right) = \frac{\sqrt{2}}{2};$$
$$\sin\left(-\frac{\pi}{4}\right) = -\frac{\sqrt{2}}{2}.$$

2. $\dfrac{s(5) - s(3)}{2} = 33$ (ft./sec.)

Average velocity = total distance/time.

3. (a) $f'(x) = \lim\limits_{h\to 0} \dfrac{[2(x+h)^2 + 1] - (2x^2 + 1)}{h}$
   $\qquad = \lim\limits_{h\to 0}(4x + 2h) = 4x;$
   (b) $f'(-2) = -8;$
   (c) $f'(0) = 0.$

Note that all terms not containing h will "cancel out" when $f(x + h) - f(x)$ is calculated. This is always true for *polynomial* functions.

4. $f'(x) = \lim\limits_{h\to 0} \dfrac{\dfrac{1}{x+h} - \dfrac{1}{x}}{h} = \lim\limits_{h\to 0}\dfrac{-h}{hx(x+h)}$
   $\qquad = \lim\limits_{h\to 0}\dfrac{-1}{x(x+h)} = \dfrac{-1}{x^2}.$
   Hence $f'(1) = -1$ and $f'(0)$ is undefined.

Since the domain of f does not contain 0, $f'(0)$ is undefined. Note that the definition of $f'(0)$ involves the evaluation of $f(0)$. That is, $f'(0) = \lim\limits_{h\to 0}\dfrac{f(0 + h) - f(0)}{h}$.

5. $f'(x) = \lim\limits_{h\to 0}\dfrac{\sqrt{x+h} - \sqrt{x}}{h}$
   $\qquad = \lim\limits_{h\to 0}\dfrac{\sqrt{x+h} - \sqrt{x}}{h} \times \dfrac{\sqrt{x+h} + \sqrt{x}}{\sqrt{x+h} + \sqrt{x}}$
   $\qquad = \lim\limits_{h\to 0}\dfrac{1}{\sqrt{x+h} + \sqrt{x}} = \dfrac{1}{2\sqrt{x}}.$

Multiplying numerator and denominator by the conjugate of $\sqrt{x+h} - \sqrt{x}$ is the key to eliminating the factor of h from the denominator. This is a "trick" that is used frequently in mathematics. The simple algebraic identity $a^2 - b^2$ is often "worth its weight in gold."

6. (a) $-6x$;
   (b) $-14x^{-3}$;
   (c) $3x^{-2/3}$.

Section 3.2: Derivatives of Elementary Functions

$$\frac{d}{dx}(ax^n) = nax^{n-1} \qquad \frac{d}{dx}(x) = 1 \qquad \frac{d}{dx}(c) = 0 \text{ for any constant } c$$

$$\frac{d}{dx}(kf(x)) = kf'(x) \qquad \frac{d}{dx}(\sin x) = \cos x \qquad \frac{d}{dx}(\cos x) = -\sin x.$$

Q1. Find $\frac{dy}{dx}$ if: (a) $y = 2x^7$ _____ (b) $y = x^{1/2}$ _____

(c) $y = 6 \cos x$ _____

Answers: (a) $14x^6$.   (b) $\frac{1}{2}x^{-1/2} = \frac{1}{2\sqrt{x}}$.   (c) $-6\sin x$,

---

 Common error: If $y = 5^x$, then $\frac{dy}{dx} = x5^{x-1}$.

Authors' Note: If $a = 1$, the first formula becomes $\frac{d}{dx}(x^n) = nx^{n-1}$. Note that x^n is a variable raised to a constant power, while 5^x is a constant raised to a variable power. The formula does not apply to 5^x.

Problem Set 3.2

Find $\frac{dy}{dx}$ for each function.

1. $y = 5x^3$

2. $y = 3x^{3/2}$

3. $f(x) = 2\sqrt{x}$

4. $y = \dfrac{\cos x}{2}$

5. $f(x) = 3 \sin x$

6. $y = 1/x$

7. $y = 3x$

8. $y = x^9 \div x^{-2}$

9. $y = -\sqrt{2}$

Solutions and Comments for Problem Set 3.2

1. $15x^2$.

2. $\frac{9}{2}x^{1/2}$.

3. $x^{-1/2}$.

4. $-\frac{1}{2}\sin x$.

5. $3 \cos x$.

8. $11x^{10}$.

6. $-x^{-2}$.

> Simplify $x^9 \div x^{-2}$ to x^{11}.

> Write $1/x$ as x^{-1}.

9. 0.

7. 3.

> The derivative of a constant is zero.

Section 3.3: Derivatives of Sums, Products, and Quotients

If u and v are differentiable functions of x, then

$$\frac{d}{dx}(u + v) = u' + v' = \frac{du}{dx} + \frac{dv}{dx}$$

$$\frac{d}{dx}(uv) = uv' + vu' = u\frac{dv}{dx} + v\frac{du}{dx}$$

$$\frac{d}{dx}\left(\frac{u}{v}\right) = \frac{vu' - uv'}{v^2} = \frac{v\dfrac{du}{dx} - u\dfrac{dv}{dx}}{v^2}$$

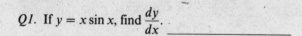

Authors' Note: The derivative of a product is *not* the product of the derivatives. The derivative of a quotient is *not* the quotient of the derivatives. For example, consider $x^2 x^5$. Since this is x^7, the derivative is $7x^6$. Note that the product of the derivative of x^2 and x^5 is $2x \times 5x^4 = 10x^5$.

Q1. If $y = x \sin x$, find $\dfrac{dy}{dx}$. _____

> Answer: $\dfrac{dy}{dx} = x \cos x + \sin x$.

Q2. Find $f'(x)$ if $f(x) = \dfrac{\cos x}{x^2}$. _____

> Answer: $f'(x) = \dfrac{-x^2 \sin x - (\cos x)(2x)}{(x^2)^2} = \dfrac{-x^2 \sin x - 2x \cos x}{x^4} = \dfrac{-(x \sin x + 2 \cos x)}{x^3}$.

Common errors: Expressing the derivative of a quotient formula as $\dfrac{vu' + uv'}{v^2}$ or as $\dfrac{uv' + vu'}{v^2}$.

Problem Set 3.3

In problems 1–11, differentiate with respect to x.

1. $y = x^5 - x^3/3 - 2x^2 + 3$

2. $y = (3x - 2)^2$

3. $f(x) = x^2 - 1/x - \sqrt{2}$

4. $g(x) = x^{3/2} - x^{1/2} - x^{-1/2}$

5. $y = x^4 - 3x^3 - x^2 - 5x - 5$

6. $f(x) = x^{-3} - x^{-2} - x^{-1} - 1$

7. $y = 2x^2 \cos x$

8. $y = \tan x$

9. $s(x) = \sec x$

10. $y = 2 \csc x$

11. $y = \dfrac{3 - x}{x^3}$

12. If $f(x) = 2x^3/3 + 5x^2/2 - 3x + 2$, find the values of x for which the tangent line to the curve is horizontal.

Solutions and Comments for Problem Set 3.3

1. $y' = 5x^4 - x^2 - 4x$.

2. $y' = 18x - 12$.

> Write $(3x - 2)^2$ as $9x^2 - 12x + 4$.

3. $f'(x) = 2x + x^{-2}$.

4. $g'(x) = \frac{3}{2}x^{1/2} - \frac{1}{2}x^{-1/2} + \frac{1}{2}x^{-3/2}$.

5. $y' = 4x^3 - 9x^2 - 2x - 5$.

6. $f'(x) = -3x^{-4} + 2x^{-3} + x^{-2}$.

> *Common error:*
> $f'(x) = -3x^{-2} + 2x^{-1} + 1$.

7. $y' = -2x^2 \sin x + 4x \cos x$.

8. $y = \dfrac{\sin x}{\cos x}$; hence
$$y' = \frac{(\cos x)(\cos x) - (\sin x)(-\sin x)}{\cos^2 x}$$
$$= \frac{1}{\cos^2 x} = \sec^2 x.$$

> $\dfrac{d}{dx}(\tan x) = \sec^2 x$ is used frequently in calculus.

9. $s(x) = \dfrac{1}{\cos x}$; hence
$$s'(x) = \frac{(\cos x)(0) - 1(-\sin x)}{\cos^2 x}$$
$$= \frac{\sin x}{\cos^2 x} = \sec x \tan x.$$

> This is used frequently in calculus.

10. $s'(x) = -2 \cot x \csc x$.

> Similar to problem #9; write $s(x) = \dfrac{2}{\sin x}$ and use quotient formula.

11. $y' = -9x^{-4} + 2x^{-3}$.

> Easiest approach is to write
> $y = 3x^{-3} - x^{-2}$.

12. $f'(x) = 2x^2 + 5x - 3 = 0$ when
    $(2x - 1)(x + 3) = 0$; that is, when $x = 1/2$ or
    $x = -3$.

> Slope of a horizontal line is 0.

Section 3.4: Derivatives of Composite Functions
(Chain Rule)

If $y = f(g(x))$, then $\dfrac{dy}{dx} = f'(g(x))g'(x)$. This is the *chain rule* for derivatives. The rule can be expressed in other forms. For instance, if $y = f(u)$ and $u = g(x)$, then $\dfrac{dy}{dx} = f'(u)g'(x) = \dfrac{dy}{du}\dfrac{du}{dx}$.

Q1. Find $\dfrac{dy}{dx}$: (a) $y = (2x - 3)^4$ _____

(b) $y = \sin^5 x$ _____

> Answers: (a) $\dfrac{dy}{dx} = 4(2x - 3)^3(2) = 8(2x - 3)^3$.
>
> (b) $\dfrac{dy}{dx} = 5\sin^4(x)\cos(x)$.

 Common error: Given $y = \sin^5 x$, $\dfrac{dy}{dx} = 5\sin^4 x$.

The formulas in Section 3.2 can now be expressed in a more meaningful form:

$$\frac{d}{dx}(au^n) = nau^{n-1}\left(\frac{du}{dx}\right) \qquad \frac{d}{dx}(\sin u) = (\cos u)\frac{du}{dx} \qquad \frac{d}{dx}(\cos u) = (-\sin u)\frac{du}{dx}$$

Q2. Find $\dfrac{dy}{dx}$: $y = \sin(x^2 - 3x)$ _____

> Answer: $\dfrac{dy}{dx} = (2x - 3)\cos(x^2 - 3x)$.

The chain rule can be extended to any number of compositions of functions.

Q3. Given $y = f(g(h(x)))$, find $\dfrac{dy}{dx}$. _____

> Answer: $\dfrac{dy}{dx} = f'(g(h(x)))g'(h(x))h'(x)$.

Authors' Note: If $y = f(u)$, $u = g(v)$, and $v = h(x)$, we can write

$$\frac{dy}{dx} = \frac{dy}{du}\frac{du}{dv}\frac{dv}{dx}.$$

Note that the right-hand side of this equation "reduces" to $\frac{dy}{dx}$.

Q4. If $y = u^2$, $u = v^3$, and $v = \cos x$, find $\frac{dy}{dx}$. _____

Answer: $\frac{dy}{dx} = (2u)(3v^2)(-\sin x) = (6v^5)(-\sin x) = -6\cos^5 x \sin x.$

Problem Set 3.4

In problems 1–6, differentiate with respect to x.

1. $y = (x^2 - 1)^4$

2. $y = \cos^3(2x)$

3. $f(x) = \dfrac{2x}{\cos 2x}$

4. $y = \left(\dfrac{2x + 1}{2x - 1}\right)^3$

5. $g(x) = x(2x - 1)^3$

6. $y = \sin(\cos x)$

In problems 7–10, find $\frac{dy}{dx}$.

7. $y = u^2$, $u = x^3 + 4x$

8. $y = u^2 - 2u - 1$, $u = \dfrac{1 + x}{1 - x}$

9. $y = 3t$, $t = u^4 + u$, $u = x^2$

10. $y = w^2 + 4w$, $w = \sin z$, $z = \pi x$

Solutions and Comments for Problem Set 3.4

1. $4(x^2 - 1)^3(2x) = 8x(x^2 - 1)^3.$

It's easy to forget about the factor of $2x$, which is the derivative of the expression inside the parentheses.

2. $3\cos^2(2x)(-\sin 2x)(2) = -6\cos^2(2x)\sin(2x).$

It might be helpful to write the function as $(\cos 2x)^3$.

3. $\dfrac{2\cos 2x + 4x\sin 2x}{\cos^2 2x}.$

Use quotient rule and chain rule.

4. $3\left(\dfrac{2x + 1}{2x - 1}\right)^2\left[\dfrac{(2x - 1)(2) - (2x + 1)(2)}{(2x - 1)^2}\right]$
$= \dfrac{-12(2x + 1)^2}{(2x - 1)^4}.$

It would be awkward to graph this function, but note that it does have a horizontal tangent at $x = -\frac{1}{2}$.

5. $x[3(2x - 1)^2(2)] + (2x - 1)^3(1)$
$= (2x - 1)^2(8x - 1).$

It is often useful to express derivatives in factored form since this is extremely helpful if one has to solve $y' = 0$.

6. $[\cos(\cos x)](-\sin x) = -\sin x \cos(\cos x)$.

> Think of $y = \sin u$, where $u = \cos x$.

7. $2u(3x^2 + 4) = 2(x^3 + 4x)(3x^2 + 4)$.

> We could write $y = (x^3 + 4x)^2$.

8. $(2u - 2)\dfrac{1 - x - (1 + x)(-1)}{(1 - x)^2} = \dfrac{8x}{(1 - x)^3}$.

9. $3(4u^3 + 1)(2x) = 3(4x^6 + 1)(2x) = 6x(4x^6 +)$.

10. $(2w + 4)(\cos z)(\pi) = \pi(\cos \pi x)(2 \sin \pi x + 4)$.

Section 3.5: Derivatives of Implicitly Defined Functions

In the form $y = f(x)$, the variable y stands alone on one side of the equation. The equation defines y *explicitly* as a function of x. If y is not isolated in an expression involving the variables x and y, then y is defined *implicitly* as a function of x.

Q1. Given that $y = f(x)$, state whether the form is implicit or explicit.

  (a) $y = 3x - 1$ _____

  (b) $3x - xy - y^2 = 0$ _____

  (c) $y = x^3 - 5x + 2$ _____

  (d) $x^4 - 2xy + 3x^2y^4 + y^5 = 0$ _____

> **Answers:** (a) explicit.   (b) implicit.   (c) explicit.   (d) implicit.

If y is defined implicitly as a function of x, one can differentiate term by term to obtain $\dfrac{dy}{dx}$. It must be remembered that y is a function of x; hence an expression such as y^2 is a composite function. Note that $\dfrac{d}{dx}(y^2) = 2y\dfrac{dy}{dx}$.

Authors' Note: Watch the variable of differentiation. Don't confuse $\dfrac{d}{dx}(y^2)$ with $\dfrac{d}{dy}(y^2) = 2y$. One might also note that if $y = f(x)$, then $y^2 = [f(x)]^2$; hence $\dfrac{d}{dx}[f(x)]^2 = 2f(x)f'(x)$ by the chain rule.

Q2. If $x^2 + y^2 + xy = 2$, find $\dfrac{dy}{dx}$. _____

> **Answer:** Differentiating implicitly, $2x + 2y\dfrac{dy}{dx} + x\dfrac{dy}{dx} + y = 0$
>
> $$\dfrac{dy}{dx}(2y + x) = -2x - y$$
>
> $$\dfrac{dy}{dx} = \dfrac{-2x - y}{2y + x}.$$

Common error: Writing $2x + 2y\dfrac{dy}{dx} + x\dfrac{dy}{dx} = 0$ as the first step in Q2; that is, overlooking the fact that the product rule must be applied to the term xy.

Q3. If $x^2 + y^2 = 4$, find: (a) $\dfrac{dy}{dx}$ _____ (b) $\dfrac{d^2y}{dx^2}$ _____

Answers: (a) $2x + 2y\dfrac{dy}{dx} = 0 \Rightarrow \dfrac{dy}{dx} = -\dfrac{x}{y}$.

(b) $\dfrac{d^2y}{dx^2} = -\dfrac{y \times 1 - x(-\frac{x}{y})}{y^2} = \dfrac{-(x^2 + y^2)}{y^3} = \dfrac{-4}{y^3}$.

Authors' Note: In Q3, more than one function is defined by the equation of the circle. The graph consists of the union of the two functions $y = \sqrt{4 - x^2}$ and $y = -\sqrt{4 - x^2}$. $\dfrac{dy}{dx}$ represents the derivative of each of these using the appropriate value for y.

Problem Set 3.5

In problems 1–5, find $\dfrac{dy}{dx}$.

1. $3x - 4y + 2xy = 3$

2. $x + \sin y = y^2$

3. $x^2 - 2xy + y^5 = 4$

4. $\dfrac{x}{y} - 2x = y$

5. $y^{1/2} - x^{1/2} = 2$

6. If $xy = 1$, find the value of $\dfrac{dy}{dx}$ at the point where $x = 1$.

7. If $x^2 - 3y^2 = 13$, find the value of $\dfrac{dy}{dx}$ at $(5, 2)$.

8. Find $\dfrac{d^2y}{dx^2}$ if (a) $y^2 = x^2 - 2$
   (b) $y^2 = 4x$

Solutions and Comments for Problem Set 3.5

1. $3 - 4\dfrac{dy}{dx} + 2x\dfrac{dy}{dx} + 2y = 0 \Rightarrow \dfrac{dy}{dx} = \dfrac{-2y - 3}{2x - 4}$.

2. $1 + (\cos y)\dfrac{dy}{dx} = 2y\dfrac{dy}{dx} \Rightarrow \dfrac{dy}{dx} = \dfrac{1}{2y - \cos y}$.

3. $2x - 2x\dfrac{dy}{dx} - 2y + 5y^4\dfrac{dy}{dx} = 0 \Rightarrow$

$\dfrac{dy}{dx} = \dfrac{2y - 2x}{5y^4 - 2x}$.

In problem #1, y can be expressed explicitly in terms of x. However, it is much easier to differentiate implicitly. Note that in most of the other problems, and specifically in #2 and #3, it would be difficult to express y explicitly.

4. $\dfrac{y - x\dfrac{dy}{dx}}{y^2} - 2 = \dfrac{dy}{dx} \Rightarrow \dfrac{dy}{dx} = \dfrac{y - 2y^2}{y^2 + x}$.

5. $\dfrac{1}{2}y^{-1/2}\dfrac{dy}{dx} - \dfrac{1}{2}x^{-1/2} = 0 \Rightarrow \dfrac{dy}{dx} = \sqrt{\dfrac{y}{x}}$.

6. $x\dfrac{dy}{dx} + y = 0 \Rightarrow \dfrac{dy}{dx} = \dfrac{-y}{x}$; at $(1, 1)$, $\dfrac{dy}{dx} = -1$.

7. $2x - 6y\dfrac{dy}{dx} = 0 \Rightarrow \dfrac{dy}{dx} = \dfrac{x}{3y}$; at $(5, 2)$, $\dfrac{dy}{dx} = \dfrac{5}{6}$.

8. (a) $\dfrac{dy}{dx} = \dfrac{x}{y}$;

$\dfrac{d^2y}{dx^2} = \dfrac{y - x\dfrac{dy}{dx}}{y^2} = \dfrac{y - \dfrac{x^2}{y}}{y^2} = \dfrac{y^2 - x^2}{y^3} = \dfrac{-2}{y^3}$.

(b) $\dfrac{dy}{dx} = \dfrac{2}{y} \Rightarrow \dfrac{d^2y}{dx^2} = \dfrac{(y)(0) - 2\dfrac{dy}{dx}}{y^2} = -\dfrac{4}{y^3}.$

> In 8(a), be aware of the original relationship between x and y. Note the simplicity of the second derivative if you remember that $y^2 - x^2 = -2$.

Section 3.6: Derivatives of Inverse Functions Including Inverse Trigonometric Functions

If $y = f(x)$ is a one-to-one differentiable function whose inverse is g, then $g'(y) = \dfrac{1}{f'(x)}$ if $f'(x) \neq 0$. Another equivalent expression for this formula is $\dfrac{dx}{dy} = \dfrac{1}{\dfrac{dy}{dx}}$.

Authors' Note: The result above is easy to prove using the chain rule. If $y = f(x)$ and $x = g(y)$, then $y = f(g(y))$. Differentiating with respect to y yields $1 = f'(g(y))g'(y) \Rightarrow 1 = f'(x)g'(y) \Rightarrow g'(y) = 1/f'(x)$.

Q1. Given $f(x) = 5x^3 - 2$, find $(f^{-1})'(3)$. _____

> **Answer:** $(f^{-1})'(y) = \dfrac{1}{15x^2}$. Since $f(1) = 3$, $(f^{-1})'(3) = \dfrac{1}{15 \times 1^2} = \dfrac{1}{15}$.

The derivatives of the inverse trigonometric functions are important in calculus. All of these formulas can be derived by implicit differentiation. For instance, if $y = \arcsin x$, then $x = \sin y$. Differentiating with respect to x yields $1 = \cos y \dfrac{dy}{dx} \Rightarrow \dfrac{dy}{dx} = \dfrac{1}{\cos y} = \dfrac{1}{\sqrt{1 - x^2}}$. (See figure.)

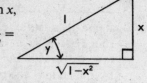

$$\frac{d}{dx}(\arcsin u) = \frac{1}{\sqrt{1 - u^2}} \cdot \frac{du}{dx} \qquad \frac{d}{dx}(\arccos u) = \frac{-1}{\sqrt{1 - u^2}} \cdot \frac{du}{dx}$$

$$\frac{d}{dx}(\arctan u) = \frac{1}{1 + u^2} \cdot \frac{du}{dx} \qquad \frac{d}{dx}(\text{arc csc } u) = \frac{-1}{u\sqrt{u^2 - 1}} \cdot \frac{du}{dx}$$

$$\frac{d}{dx}(\text{arc sec } u) = \frac{1}{u\sqrt{u^2 - 1}} \cdot \frac{du}{dx} \qquad \frac{d}{dx}(\text{arc cot } u) = \frac{-1}{1 + u^2} \cdot \frac{du}{dx}$$

Q2. Find $\dfrac{dy}{dx}$ if $y = \arcsin(2x)$. _____

Answer: $\dfrac{dy}{dx} = \dfrac{2}{\sqrt{1 - 4x^2}}$.

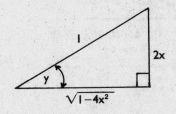

Problem Set 3.6

In problems 1–4, find the derivative of g, given that $g = f^{-1}$.

1. $f(x) = x^2,\ x \geq 0$

2. $f(x) = \sqrt{x}$

3. $f(x) = x^3 + 1$

4. $f(x) = \dfrac{1}{x + 1}$

In problems 5–8, find the derivative.

5. $y = \arcsin(3x)$

6. $f(x) = x \arctan x$

7. $g(x) = \arccos \tfrac{1}{2}$

8. $y = (\arcsin x)^{-1}$

9. If $f(x) = x^3 + x$, find $(f^{-1})'(10)$.

10. Derive the formula for $\dfrac{d}{dx}(\arctan x)$ using implicit differentiation.

Solutions and Comments for Problem Set 3.6

1. $g'(y) = \dfrac{1}{2x} = \dfrac{1}{2\sqrt{y}}$; hence $g'(x) = \dfrac{1}{2\sqrt{x}}$.

> Remember that g' can be thought of as a rule. Here, g' says "take a positive number, take its principal square root, multiply it by 2, then take the reciprocal."
>
> Note that $g'(y) = \dfrac{1}{2\sqrt{y}}$ is the same "rule" as $g'(x) = \dfrac{1}{2\sqrt{x}}$. Don't be confused by choice of variable.

2. $g'(y) = \dfrac{1}{\frac{1}{2}x^{-1/2}} = 2\sqrt{x} = 2y$; hence $g'(x) = 2x$.

3. $g'(y) = \dfrac{1}{3x^2} = \dfrac{1}{3(y - 1)^{2/3}}$.

> One could, of course, find the inverse function and then differentiate. In this case, $f^{-1}(x) = g(x) = (x - 1)^{1/3}$ and $g'(x) = \tfrac{1}{3}(x - 1)^{-2/3}$, the identical "rule" obtained above.

4. $g'(y) = \dfrac{1}{-(x + 1)^{-2}} = -(x + 1)^2 = -\dfrac{1}{y^2}$.

5. $\dfrac{3}{\sqrt{1 - 9x^2}}$.

> Apply formula from reading section and remember chain rule.

6. $x \times \dfrac{1}{1 + x^2} + 1 \times \arctan x$

$= \dfrac{x}{1 + x^2} + \arctan x$.

7. $g'(x) = 0$.

> Note that $\arccos \frac{1}{2}$ is a constant and the derivative of a constant is zero.

$$f(2) = 10 \text{ and } f'(x) = 3x^2 + 1.$$

10. Differentiating $x = \tan y$ with respect to x yields $1 = \sec^2 y \dfrac{dy}{dx}$. Hence $\dfrac{dy}{dx} = \dfrac{1}{\sec^2 y} = \dfrac{1}{1 + x^2}$. (See figure below.)

8. $-1(\arcsin x)^{-2} \times \dfrac{1}{\sqrt{1 - x^2}}$

$= \dfrac{-1}{\sqrt{1 - x^2}(\arcsin x)^2}$.

> If $y = x^{-1}$, then $y' = -1 \times x^{-2}$.

9. $(f^{-1})'(10) = \dfrac{1}{f'(2)} = \dfrac{1}{13}$.

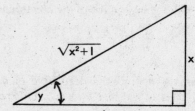

Section 3.7: Derivatives of Logarithmic and Exponential Functions

Remember: $e = \lim\limits_{n \to \infty} \left(1 + \dfrac{1}{n}\right)^n$, which is approximately 2.71828.

> (1) $\dfrac{d}{dx}(\log_b x) = \dfrac{1}{x} \log_b e$ (2) $\dfrac{d}{dx}(\log_b u) = \dfrac{1}{u}(\log_b e)\dfrac{du}{dx}$

Q1. Find $\dfrac{dy}{dx}$: (a) $y = \log_{10} x$ _____   (b) $y = \log_3 7x$ _____

> Answers: (a) $\dfrac{1}{x} \log_{10} e$.   (b) $\dfrac{1}{7x}(\log_3 e)\dfrac{d}{dx}(7x) = \dfrac{1}{7x}(\log_3 e) \times 7 = \dfrac{1}{x}\log_3 e$.

Authors' Notes: Alternate approach for (b); $y = \log_3 7 + \log_3 x$; hence $\dfrac{dy}{dx} = 0 + \dfrac{1}{x}\log_3 e$.

If $u = x$ in formula (2), the result is equivalent to formula (1).

If the base b happens to be e, then $\dfrac{d}{dx}\log_e x = \dfrac{1}{x}\log_e e = \dfrac{1}{x} \times 1 = \dfrac{1}{x}$. In other words,

> (3) $\dfrac{d}{dx}(\ln x) = \dfrac{1}{x}$ (4) $\dfrac{d}{dx}(\ln u) = \dfrac{1}{u} \times \dfrac{du}{dx}$

Authors' Note: Choosing e as the base for a logarithmic system greatly simplifies work involving derivatives of logarithm.

Q2. Find $\dfrac{dy}{dx}$: (a) $y = \ln 6x$ _____ (b) $y = \ln(\sin x)$ _____

Answers: (a) $\dfrac{1}{6x} \times \dfrac{d}{dx}(6x) = \dfrac{1}{6x}(6) = \dfrac{1}{x}$. (b) $\dfrac{1}{\sin x} \times (\cos x) = \dfrac{\cos x}{\sin x} = \cot x$.

Authors' Note: The domain of the function in (b) would include only those values of x for which $0 < \sin x \le 1$.

If $y = a^x$, then $\ln y = \ln a^x = x \ln a$. Differentiating, one obtains $\dfrac{1}{y} \times \dfrac{dy}{dx} = \ln a \Rightarrow \dfrac{dy}{dx} = y \ln a = a^x \ln a$. In other words,

(5) $\dfrac{d}{dx}(a^x) = a^x \ln a$

From (5), one can easily deduce

(6) $\dfrac{d}{dx}(e^x) = e^x$ (7) $\dfrac{d}{dx}(e^u) = e^u \dfrac{du}{dx}$

Q3. Find $\dfrac{dy}{dx}$: (a) $y = 5^x$ _____ (b) $y = e^{7x^3}$ _____

Answers: (a) $5^x \ln 5$. (b) $e^{7x^3} \times \dfrac{d}{dx}(7x^3) = 21x^2 e^{7x^3}$.

 Common errors: If $y = 5^x$, then $\dfrac{dy}{dx} = x5^{x-1}$. If $y = e^{7x^3}$, then $\dfrac{dy}{dx} = e^{7x^3}$.

Q4. If $y = e^{\sin^2 x}$, find $\dfrac{dy}{dx}$. _____

Answer: $e^{\sin^2 x} \dfrac{d}{dx}(\sin^2 x) = e^{\sin^2 x} 2 \sin x \dfrac{d}{dx}(\sin x) = e^{\sin^2 x} 2 \sin x \cos x$, which can be written $e^{\sin^2 x} \sin 2x$.

 Common error: In Q4, $\dfrac{dy}{dx} = e^{\sin^2 x} 2 \sin x$.

In many physical applications, functions arise that are combinations of e^x and e^{-x}. Some of these functions have special names:

Hyperbolic sine: $\sinh x = \frac{1}{2}(e^x - e^{-x})$ Hyperbolic cosine: $\cosh x = \frac{1}{2}(e^x + e^{-x})$

With these definitions it is easy to prove that

$$\frac{d}{dx}(\sinh u) = \cosh u \frac{du}{dx} \qquad \frac{d}{dx}(\cosh u) = \sinh u \frac{du}{dx}$$

The graphs of the hyperbolic sine and the hyperbolic cosine are shown below.

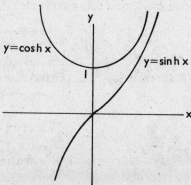

Authors' Note: From the definitions, one can show that

$$\cosh^2 x - \sinh^2 x = 1.$$

This is one of the reasons why the functions are called hyperbolic functions. The form $x^2 - y^2 = 1$ is that of a hyperbola.

There are four other hyperbolic functions. They are given here for reference:

$$\tanh x = \frac{\sinh x}{\cosh x} = \frac{e^x - e^{-x}}{e^x + e^{-x}} \qquad \coth x = \frac{1}{\tanh x} = \frac{e^x + e^{-x}}{e^x - e^{-x}}$$

$$\operatorname{sech} x = \frac{1}{\cosh x} = \frac{2}{e^x + e^{-x}} \qquad \operatorname{csch} x = \frac{1}{\sinh x} = \frac{2}{e^x - e^{-x}}$$

Note that $\coth x$ and $\operatorname{csch} x$ are not defined at $x = 0$. The domain for the other hyperbolic functions is $(-\infty, \infty)$.

Problem Set 3.7

1. Differentiate the following:
   (a) $\log_5 4x$
   (b) $\ln(x^2 + 7x)$
   (c) $\ln[(x^2 + 7x)^{35}]$
   (d) e^{x^2}
   (e) $\dfrac{e^x + e^{-x}}{2}$
   (f) $\ln \sqrt[3]{x^4 + x^2}$
   (g) 2^x
   (h) $(\ln 2x)^3$
   (i) $\ln \dfrac{1 + x^2}{2 + x^2}$

2. Differentiate
   (a) $\ln(\sin^2 x + \cos^2 x)$
   (b) $x \ln x$

(c) $x^2 e^x$

(d) $e^x \sin x$

3. If $y = e^{\cos x}$, find all values of x for which $\dfrac{dy}{dx} = 0$.

4. If $f(x) = e^{-x}$, find all solutions for $f'(x) = -1$.

In problems 5–7, find $f'(x)$.

5. $f(x) = \sinh(2x + 3)$

6. $f(x) = \cosh(\tfrac{1}{2}x)$

7. $f(x) = [\sinh(4x)][\cosh(3x)]$

8. Show that $\dfrac{d}{dx}(\coth x) = -\operatorname{csch}^2 x$.

Solutions and Comments for Problem Set 3.7

1. (a) $\dfrac{1}{x}\log_5 e$;

(b) $\dfrac{2x + 7}{x^2 + 7x}$;

(c) $\dfrac{35(2x + 7)}{x^2 + 7x}$;

(d) $2xe^{x^2}$;

(e) $\dfrac{e^x - e^{-x}}{2}$;

(f) $\dfrac{1}{3}\left(\dfrac{4x^3 + 2x}{x^4 + x^2}\right)$;

(g) $2^x \ln 2$;

(h) $3(\ln 2x)^2 \dfrac{d}{dx}(\ln 2x) = \dfrac{3(\ln 2x)^2}{x}$;

(i) $y = \ln(1 + x^2) - \ln(2 + x^2)$; hence

$\dfrac{dy}{dx} = \dfrac{1}{1 + x^2}(2x) - \dfrac{1}{2 + x^2}(2x)$

$= \dfrac{2x}{(1 + x^2)(2 + x^2)}$.

> Note that $\dfrac{d}{dx}\log_b(nx) = \dfrac{1}{nx}(\log_b e)(n) = \dfrac{1}{x}\log_b e$. That is, the result is independent of n. In (c), writing the expression as $35\ln(x^2 + 7x)$ makes differentiation relatively simple; compare to (b). In a similar manner, (f) can be written as $\tfrac{1}{3}\ln(x^4 + x^2)$. In (i), the logarithmic property $\log_b\left(\dfrac{a}{c}\right) = \log_b a - \log_b c$ makes the derivative calculation quite easy.

2. (a) 0;

(b) $x \times \dfrac{1}{x} + (\ln x) \times 1 = 1 + \ln x$;

(c) $x^2 e^x + e^x(2x) = e^x(x^2 + 2x)$;

(d) $e^x \cos x + (\sin x)e^x = e^x(\cos x + \sin x)$.

> Note that (a) is just $\ln 1$ and the derivative of any constant is 0. The product rule for derivatives is used in parts (b), (c), and (d).

3. $\dfrac{dy}{dx} = e^{\cos x}(-\sin x) = 0$ when $\sin x = 0$. Hence $\dfrac{dy}{dx} = 0$ when $x = n\pi$, where n is any integer.

> Note that e^t is never zero for any t.

4. $f'(x) = -e^{-x}$. If $-e^{-x} = -1$, then $e^{-x} = 1$ and $x = 0$ is the only solution.

5. $2\cosh(2x + 3)$.

6. $\tfrac{1}{2}\sinh(\tfrac{1}{2}x)$.

7. $[\sinh(4x)][3\sinh(3x)]$
  $+ [\cosh(3x)][4\cosh(4x)]$
  $= 3\sinh(4x)\sinh(3x) + 4\cosh(3x)\cosh(4x)$.

> Problems 5–7 can be done using a direct application of the formulas in the reading section.

8. $\dfrac{d}{dx}\left(\dfrac{e^x + e^{-x}}{e^x - e^{-x}}\right)$

$= \dfrac{(e^x - e^{-x})(e^x - e^{-x}) - (e^x + e^{-x})(e^x + e^{-x})}{(e^x - e^{-x})^2}$

$= \dfrac{-4}{(e^x - e^{-x})^2} = -\operatorname{csch}^2 x$.

> Using $a^2 - b^2 = (a + b)(a - b)$ saves considerable computation in the numerator.

Section 3.8: Derivatives of Higher Order

If $y = f(x)$, then the derivative of f has various notations, including $\dfrac{dy}{dx}$, y', $\dfrac{d}{dx} f(x)$ and $f'(x)$. The *second derivative* is the derivative of the first derivative: There are various notations for it, including

$$\frac{d}{dx}\left(\frac{dy}{dx}\right), \quad \frac{d^2 y}{dx^2}, \quad \frac{d^2}{dx^2} f(x), \quad y'' \quad \text{and} \quad f''(x).$$

In general, the notations for the *n-th derivative* are

$$\frac{d^n y}{dx^n}, \quad \frac{d^n}{dx^n} f(x), \quad y^{(n)} \quad \text{and} \quad f^{(n)}(x).$$

Authors' Note: Don't confuse $\dfrac{d^n y}{dx^n}$ with $\left(\dfrac{dy}{dx}\right)^n$. For instance, if $y = x^3$, we have $\dfrac{dy}{dx} = 3x^2$. Then $\dfrac{d^2 y}{dx^2} = 6x$ and $\left(\dfrac{dy}{dx}\right)^2 = (3x^2)^2 = 9x^4$. In the same vein, don't confuse $f^{(n)}(x)$ with $f^n(x) = [f(x)]^n$.

Q1. If $f(x) = x^3 + \sin x + e^{2x}$, find:

    (a) $f'(x)$ _____

    (b) $f''(x)$ _____

    (c) $f'''(x)$ _____

    (d) $f^{(4)}(x)$ _____

> Answers: (a) $f'(x) = 3x^2 + \cos x + 2e^{2x}$.   (b) $f''(x) = 6x - \sin x + 4e^{2x}$.
> (c) $f'''(x) = 6 - \cos x + 8e^{2x}$.   (d) $f^{(4)}(x) = \sin x + 16e^{2x}$.

Problem Set 3.8

1. Find the first and second derivatives for
   (a) $x^2 + 8x - 17$
   (b) $\cos e^x$
   (c) $\dfrac{x}{1 + x}$
   (d) $x \ln x$
   (e) e^{-x}
   (f) $\sin 2x$
   (g) x^{-1}
   (h) $\tan x$

2. If $y = \sin x$, show that $y^{(n)} = y^{(n+4)}$ for any positive integer n.

3. If $g(x) = e^x$, find $g^{(7)}(0)$ and $g^{(19)}(1)$.

4. If $w(x) = \dfrac{x}{x^2 + 1}$, find $w(2)$, $w'(2)$ and $w''(2)$.

5. If $y = \sin x + \cos x$, show that $y'' + y = 0$.

6. If $y = ae^{-x} + be^x$, show that $y^{(4)} - y = 0$.

Solutions and Comments for Problem Set 3.8

1. (a) $y' = 2x + 8$, $y'' = 2$;
   (b) $y' = -e^x \sin e^x$, $y'' = -e^{2x} \cos e^x - e^x \sin e^x$;
   (c) $y' = (1 + x)^{-2}$, $y'' = -2(1 + x)^{-3}$;
   (d) $y' = 1 + \ln x$, $y'' = \dfrac{1}{x}$;

   (e) $y' = -e^{-x}$, $y'' = e^{-x}$;
   (f) $y' = 2 \cos 2x$, $y'' = -4 \sin 2x$;
   (g) $y' = -x^{-2}$, $y'' = 2x^{-3}$;
   (h) $y' = \sec^2 x$, $y'' = 2 \tan x \sec^2 x$.

Remember that there are equivalent forms for some of the answers. For instance, in (h), $y'' = \dfrac{2\sin x}{\cos^3 x}$ is an equivalent answer. If one forgets the derivative of the tangent function, one can write $y = \dfrac{\sin x}{\cos x}$; using the quotient formula, one easily obtains $y' = \dfrac{1}{\cos^2 x}$, which can be written $(\cos x)^{-2}$. Applying the chain rule, $y'' = -2(\cos x)^{-3}(-\sin x)$.

2. $y = \sin x$, $y' = \cos x$, $y'' = -\sin x$, $y''' = -\cos x$, $y^{(4)} = \sin x$, $y^{(5)} = \cos x$, etc.

Using the $y^{(n)}$ notation, $y^{(1)}$ would be y', $y^{(2)}$ would be y'', etc.

3. $g^{(7)}(0) = e^0 = 1$, $g^{(19)}(1) = e^1 = e$.

For any positive integer n, $\dfrac{d^n}{dx^n}(e^x) = e^x$.

4. $w(2) = \dfrac{2}{5}$; $w'(x) = \dfrac{1 - x^2}{(x^2 + 1)^2}$; hence $w'(2) = \dfrac{-3}{25}$. $w''(x) = \dfrac{2x(x^2 - 3)}{(x^2 + 1)^3}$; hence $w''(2) = \dfrac{4}{125}$.

Instead of using the quotient formula, one could write $w(x) = x(x^2 + 1)^{-1}$ and use the product formula.

5. $y' = \cos x - \sin x$, $y'' = -\sin x - \cos x$. Hence $y'' = -y$.

Also note that $y^{(4)} = y$.

6. $y' = -ae^{-x} + be^x$, $y'' = ae^{-x} + be^x$, $y''' = -ae^{-x} + be^x$, $y^{(4)} = ae^{-x} + be^x$. Hence $y^{(4)} = y$.

See the comment for problem #5. Note that both $y = \sin x + \cos x$ and $y = ae^{-x} + be^x$ are solutions to the differential equation $y^{(4)} - y = 0$.

Section 3.9: Rolle's Theorem and the Mean Value Theorem

Rolle's Theorem: If f is a function that is continuous on $[a, b]$ and differentiable on (a, b) such that $f(a) = 0$ and $f(b) = 0$, then there is a number $x_1 \in (a, b)$ such that $f'(x_1) = 0$.

Authors' Note: From a geometric standpoint, Rolle's Theorem should be relatively obvious. Basically, given the premise of the theorem, there must be at least one horizontal tangent to the graph of the function between $x = a$ and $x = b$. Another interpretation: Between the roots $x = a$ and $x = b$ of $f(x) = 0$, there must be a maximum or minimum value for $y = f(x)$. The figures on the right and at the top of page 84 demonstrate possible situations.

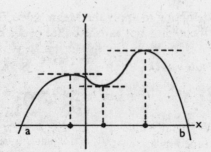

Q1. Given $f(x) = x^2 + 4x - 21$. Show that $f(-7) = 0$, $f(3) = 0$ and find all points $x_1 \in (-7, 3)$ such that $f'(x_1) = 0$. _____

> Answer: $f(-7) = (-7)^2 + 4(-7) - 21 = 49 - 28 - 21 = 0$;
> $f(3) = 3^2 + 4(3) - 21 = 9 + 12 - 21 = 0$;
> $f'(x) = 2x + 4 \Rightarrow f'(x) = 0$ when $x = -2$.

Mean Value Theorem: If f is a function that is continuous on $[a, b]$ and differentiable on (a, b) then there is at least one number $x_1 \in (a, b)$ such that $\dfrac{f(b) - f(a)}{b - a} = f'(x_1)$.

Authors' Note: Like Rolle's Theorem, which is really a corollary of this theorem, the Mean Value Theorem is geometrically obvious, as illustrated in the figures on the right. Note that $\dfrac{f(b) - f(a)}{b - a}$ is just the slope of the line AB joining $(a, f(a))$ and $(b, f(b))$. Given the premise of the Mean Value Theorem, there must be at least one line tangent to the curve that has the same slope as line AB.

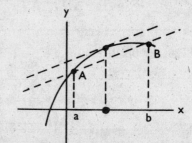

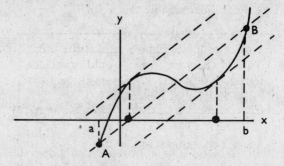

Q2. Given $f(x) = x^3$, find $x_1 \in (1, 4)$ such that $\dfrac{f(4) - f(1)}{4 - 1} = f'(x_1)$. _____

> Answer: $\dfrac{f(4) - f(1)}{4 - 1} = \dfrac{64 - 1}{3} = 21$ and $f'(x) = 3x^2$. If $f'(x_1) = 0$, then $3x_1^2 = 21 \Rightarrow x_1 = \sqrt{7}$.
> (Note that $-\sqrt{7}$ is not in the interval $(1, 4)$.)

Common error: Attempting to apply the Mean Value Theorem when the hypothesis of the theorem is not satisfied. (See problem #1 in problem set.)

Problem Set 3.9

In problems 1–4, explain why the Mean Value Theorem cannot be applied to the function on the specified interval.

1. $f(x) = x^{2/3}$ on $[-1, 8]$

2. $f(x) = \tan x$ on $\left[\dfrac{\pi}{4}, \dfrac{3\pi}{4}\right]$

3. $f(x) = |x|$ on $[-2, 2]$

4. $f(x) = x^{-1}$ on $[-2, 5]$

In problems 5–10, find all possible values of x_1 such that $\dfrac{f(b) - f(a)}{b - a} = f'(x_1)$.

5. $f(x) = x^2$, $a = 1$, $b = 4$

6. $f(x) = x^{1/2}$, $a = 1$, $b = 9$

7. $f(x) = 1/x$, $a = 1$, $b = 4$

8. $f(x) = x^4$, $a = 1$, $b = 2$

9. $f(x) = e^x$, $a = 0$, $b = 1$

10. $f(x) = x^2 - 9x + 18$, $a = 3$, $b = 6$

11. $f(x) = \sin x$ is continuous and differentiable on $[0, 1]$. Use the Mean Value Theorem to prove that if $0 < x < 1$, then $\sin x < x$.

Solutions and Comments for Problem Set 3.9

1. f is not differentiable at $x = 0$.

2. f is not continuous at $x = \frac{\pi}{2}$.

3. f is not differentiable at $x = 0$.

Note that if one blindly applies the Mean Value Theorem, one would conclude (erroneously) that there exists an $x_1 \in (-2, 2)$ such that $f'(x_1) = 0$.

4. f is not continuous at $x = 0$.

5. $x_1 = 5/2$.

6. $x_1 = 4$.

7. $x_1 = 2$.

8. $x_1 = \sqrt[3]{\dfrac{15}{4}} = \dfrac{\sqrt[3]{30}}{2}$.

9. $\dfrac{e^1 - e^0}{1 - 0} = f'(x_1) = e^{x_1}$; hence

$$e^{x_1} = e - 1 \Rightarrow x_1 = \ln(e - 1).$$

10. $x_1 = 4.5$.

Note that $f(3) = f(6) = 0$; hence we have an application of Rolle's Theorem.

11. $f(x) = \sin x \Rightarrow f'(x) = \cos x$. Let $a = 0$ and $b = x$. Then the Mean Value Theorem guarantees an $x_1 \in (0, x)$ such that $\dfrac{\sin x - \sin 0}{x - 0} = \cos x_1$. Now $\sin 0 = 0$ and on $(0, 1)$, $\cos x_1 < 1$. Hence $\dfrac{\sin x}{x} < 1 \Rightarrow \sin x < x$.

Note that $\sin x < x$ is not universally true. For instance, if $x < -1$, the statement is false.

Section 3.10: The Relationship Between Differentiability and Continuity

Theorem: Given a function f, if f has a derivative at $x = x_0$, then f is continuous at $x = x_0$.

Authors' Note: The proof of this theorem is relatively simple. If $f'(x_0)$ exists, then clearly $f(x_0)$ is defined. Also, one can write

$$f(x_0 + h) = \frac{f(x_0 + h) - f(x_0)}{h} \times h + f(x_0).$$

Thus,

$$\lim_{h \to 0} f(x_0 + h) = \lim_{h \to 0} \left[\frac{f(x_0 + h) - f(x_0)}{h} \times h + f(x_0) \right]$$

$$= \lim_{h \to 0} \frac{f(x_0 + h) - f(x_0)}{h} \times \lim_{h \to 0} h + \lim_{h \to 0} f(x_0)$$

$$= f'(x_0) \times 0 + f(x_0)$$

$$= f(x_0).$$

The converse of the theorem above is not true; that is, a function continuous at $x = x_0$ does not necessarily have a derivative at $x = x_0$.

Q1. Consider $f(x) = |x|$.

    (a) Is f continuous at $x = 0$? _____

    (b) Is f differentiable at $x = 0$? _____

Answers: (a) f is continuous at $x = 0$; that is, $f(0) = 0$ and $\lim_{x \to 0} f(x) = 0$.

    (b) f is not differentiable at $x = 0$. Note that if $h > 0$, then

$$\lim_{h \to 0} \frac{|0 + h| - 0}{h} = \lim_{h \to 0} \frac{h}{h} = \lim_{h \to 0} 1 = 1 \text{ and if } h < 0, \text{ then}$$

$$\lim_{h \to 0} \frac{|0 + h| - 0}{h} = \lim_{h \to 0} \frac{-h}{h} = \lim_{h \to 0} (-1) = -1; \text{ that is,}$$

$$\lim_{h \to 0} \frac{|0 + h| - 0}{h} \text{ does not exist.}$$

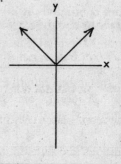

Authors' Note: There are three interesting functions that contrast differentiability and continuity.

$$f(x) = \begin{cases} \sin\frac{1}{x} & \text{if } x \neq 0 \\ 0 & \text{if } x = 0 \end{cases} \text{ is not continuous at } x = 0 \text{ and not differentiable at } x = 0.$$

$$g(x) = \begin{cases} x \sin\frac{1}{x} & \text{if } x \neq 0 \\ 0 & \text{if } x = 0 \end{cases} \text{ is continuous at } x = 0, \text{ but not differentiable there.}$$

$$k(x) = \begin{cases} x^2 \sin\frac{1}{x} & \text{if } x \neq 0 \\ 0 & \text{if } x = 0 \end{cases} \text{ is differentiable and continuous at } x = 0.$$

Problem Set 3.10

In problems 1–5, determine if the function is continuous at $x = 0$ and if it has a derivative at $x = 0$.

1. $f(x) = x^2$

2. $f(x) = \dfrac{|x|}{x}$

3. $f(x) = x^{1/3}$

4. $g(x) = \ln x$

5. $w(x) = \dfrac{x^2 - 9}{x - 3}$

6. Consider $f(x) = |x - 2|$.
   (a) Is f continuous at $x = 2$?
   (b) Is f differentiable at $x = 2$?

7. From the graphs, determine if the function is
   (i) continuous at $x = c$;
   (ii) differentiable at $x = c$.

(a)   (b)   (c)   (d)

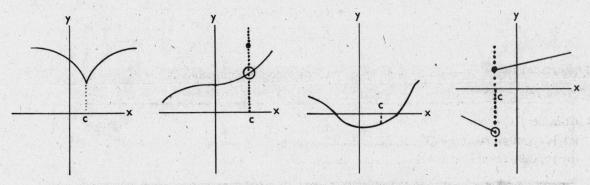

Solutions and Comments for Problem Set 3.10

1. Continuous and differentiable.

2. Neither continuous nor differentiable.

3. Continuous, but not differentiable.

4. Neither continuous nor differentiable.

5. Continuous and differentiable.

> The functions in (2) and (4) are not defined at $x = 0$. In (3), the tangent to the curve is vertical at $x = 0$. In (5), the function is neither continuous nor differentiable at $x = 3$.

6. (a) yes;
   (b) no.

> The graph of the function is congruent to $y = |x|$ (see Q1 in reading section), but shifted two units to the right.

7. (a) continuous, but not differentiable;
   (b) neither continuous nor differentiable;
   (c) continuous and differentiable;
   (d) neither continuous nor differentiable.

> The derivative, if it exists, represents the slope of a tangent line at a point on the curve. Note that in (a), there would not be a unique tangent line to the curve at $x = c$. Contrast this with the graph in part (c). Loosely speaking, a function would not have a derivative at a "sharp" point. "Smoothness" and existence of a derivative are somewhat synonymous.

Section 3.11: L'Hopital's Rule for Quotient Indeterminate Forms

The statements in the shaded areas are true if a is a real number or if a is replaced by ∞.

If $\lim\limits_{x \to a} f(x) = 0$ and $\lim\limits_{x \to a} g(x) = 0$, then $\lim\limits_{x \to a} \dfrac{f(x)}{g(x)}$ is called an *indeterminate of the form* $\dfrac{0}{0}$. If $\lim\limits_{x \to a} f(x) = \infty$ (or $-\infty$) and $\lim\limits_{x \to a} g(x) = \infty$ (or $-\infty$), then $\lim\limits_{x \to a} \dfrac{f(x)}{g(x)}$ is called an *indeterminate of the form* $\dfrac{\infty}{\infty}$.

The following rule is extremely useful in evaluating certain types of limits:

L'Hopital's Rule: If $\lim\limits_{x \to a} \dfrac{f(x)}{g(x)}$ is an indeterminate of the form $\dfrac{0}{0}$ or $\dfrac{\infty}{\infty}$, and if $\lim\limits_{x \to a} \dfrac{f'(x)}{g'(x)}$ exists, then $\lim\limits_{x \to a} \dfrac{f(x)}{g(x)} = \lim\limits_{x \to a} \dfrac{f'(x)}{g'(x)}$.

Q1. Find: (a) $\lim\limits_{x \to 1} \dfrac{\ln x}{x - 1}$ _____ (b) $\lim\limits_{x \to 0} \dfrac{\ln x}{x^{-1}}$ _____

Answers: (a) This is an indeterminate of the form $\dfrac{0}{0}$; hence, $\lim\limits_{x \to 1} \dfrac{\ln x}{x - 1} = \lim\limits_{x \to 1} \dfrac{\dfrac{d}{dx}(\ln x)}{\dfrac{d}{dx}(x - 1)} =$

$\lim\limits_{x \to 1} \dfrac{\dfrac{1}{x}}{1} = 1$.

(b) This is an indeterminate of the form $\dfrac{\infty}{\infty}$; hence $\lim\limits_{x \to 0} \dfrac{\ln x}{x^{-1}} = \lim\limits_{x \to 0} \dfrac{\dfrac{d}{dx}(\ln x)}{\dfrac{d}{dx}(x^{-1})} = \lim\limits_{x \to 0} \dfrac{\dfrac{1}{x}}{-\dfrac{1}{x^2}}$

$= \lim\limits_{x \to 0} (-x) = 0$.

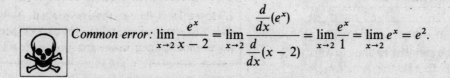

Common error: $\lim\limits_{x \to 2} \dfrac{e^x}{x - 2} = \lim\limits_{x \to 2} \dfrac{\dfrac{d}{dx}(e^x)}{\dfrac{d}{dx}(x - 2)} = \lim\limits_{x \to 2} \dfrac{e^x}{1} = \lim\limits_{x \to 2} e^x = e^2.$

Authors' Note: $\lim\limits_{x \to 2} \dfrac{e^x}{x - 2}$ is not an indeterminate form, hence L'Hopital's Rule cannot be applied. The stated limit does not exist (and hence is not e^2). Be alert to the oft-overlooked fact that L'Hopital's Rule does not guarantee the existence of a limit for an indeterminate form.

Problem Set 3.11

Find the indicated limits.

1. $\lim\limits_{x \to 0} \dfrac{e^x - 1}{x}$

2. $\lim\limits_{w \to 0} \dfrac{\sin w}{w}$

3. $\lim\limits_{x \to \infty} \dfrac{\ln x^2}{x}$

4. $\lim\limits_{x \to 1} \dfrac{\ln x}{x^2 - x}$

5. $\lim\limits_{x \to -1} \dfrac{\sin \pi x}{1 + x}$

6. $\lim\limits_{x \to 2} \dfrac{x^3 - 8}{x^2 - 4}$

7. $\lim\limits_{x \to 0} \dfrac{\ln x}{\cot x}$

8. $\lim\limits_{x \to 4} \dfrac{x^2 + 3}{x + 2}$

9. $\lim\limits_{x \to \infty} \dfrac{5.4x + 47}{-2.7x + 678}$

10. $\lim\limits_{x \to 0} \dfrac{e^x + e^{-x} - 2}{3x^2}$

11. $\lim\limits_{h \to 0} \dfrac{(4 + h)^3 - 4^3}{h}$

12. $\lim\limits_{x \to \pi/2} (\sec x - \tan x)$

Solutions and Comments for Problem Set 3.11

1. $\lim\limits_{x \to 0} \dfrac{e^x}{1} = 1.$

2. $\lim\limits_{w \to 0} \dfrac{\cos w}{1} = 1.$

> Relate this limit to Section 2.3.

3. $\lim\limits_{x \to \infty} \dfrac{\frac{2}{x}}{1} = \lim\limits_{x \to \infty} \dfrac{2}{x} = 0.$

> Remember that $\ln x^2 = 2 \ln x$.

4. $\lim\limits_{x \to 1} \dfrac{\frac{1}{x}}{2x - 1} = 1.$

5. $\lim\limits_{x \to -1} \dfrac{\pi \cos \pi x}{1} = \pi \cos(-\pi) = -\pi.$

6. $\lim\limits_{x \to 2} \dfrac{3x^2}{2x} = \dfrac{12}{4} = 3.$

7. $\lim\limits_{x \to 0} \dfrac{\frac{1}{x}}{-\csc^2 x} = \lim\limits_{x \to 0} \dfrac{-\sin^2 x}{x}$

$= \lim\limits_{x \to 0} \left[\dfrac{\sin x}{x}(-\sin x) \right] = 1 \times 0 = 0.$

> An alternate approach would be to apply L'Hopital's Rule to $\lim\limits_{x \to 0} \dfrac{-\sin^2 x}{x}$ which is an indeterminate of the form $\frac{0}{0}$.

8. $19/6.$

> L'Hopital's Rule cannot be applied. You didn't attempt to apply it, did you??? (See the common error in the reading section.)

9. $\lim\limits_{x \to \infty} \dfrac{5.4}{-2.7} = -2.$

10. $\lim\limits_{x \to 0} \dfrac{e^x - e^{-x}}{6x} = \lim\limits_{x \to 0} \dfrac{e^x + e^{-x}}{6} = \dfrac{2}{6} = \dfrac{1}{3}.$

L'Hopital's Rule was applied twice. Note that the first application produced an indeterminate of the form $\frac{0}{0}$.

11. $\lim\limits_{h \to 0} \dfrac{3(4 + h)^2}{1} = 3 \times 4^2 = 48.$

This expression represents the derivative of $y = x^3$ evaluated at $x = 4$.

12. $\sec x - \tan x = \dfrac{1}{\cos x} - \dfrac{\sin x}{\cos x} = \dfrac{1 - \sin x}{\cos x}.$

Hence,

$$\lim_{x \to \pi/2} \frac{1 - \sin x}{\cos x} = \lim_{x \to \pi/2} \frac{-\cos x}{-\sin x} = \frac{0}{-1} = 0.$$

The expressed limit is not in indeterminate form. However, it can be rewritten so that L'Hopital's Rule can be applied.

*Section 3.11A: Other Indeterminate Forms

The indeterminate form $0 \times \infty$ can be "reduced" to $\dfrac{0}{0}$ or $\dfrac{\infty}{\infty}$ so that L'Hopital's Rule can be applied.

Q1. Find $\lim\limits_{x \to 0} (x^3 \ln x).$ _____

Answer: Writing $\lim\limits_{x \to 0} (x^3 \ln x)$ as $\lim\limits_{x \to 0} \dfrac{\ln x}{\dfrac{1}{x^3}}$, one obtains the indeterminate form $\dfrac{-\infty}{\infty}$. Using

L'Hopital's Rule, the above expression is equal to $\lim\limits_{x \to 0} \dfrac{\dfrac{1}{x}}{\dfrac{-3}{x^4}} = \lim\limits_{x \to 0} \left(\dfrac{-x^3}{3} \right) = 0.$

Common error: $\lim\limits_{x \to 0} (x^3 \ln x) = 0 \times (-\infty) = 0.$

Authors' Note: If n is any real number, then $0 \times n = 0$. However ∞ is not a real number. Hence the expression $0 \times \infty$ would not necessarily represent 0. It might be beneficial to look at Q3 in Section 2.1. All three parts of this problem represent a $0 \times \infty$ situation.

The indeterminate forms 0^0, ∞^0, and 1^∞ can be handled by first taking logarithms of the form under consideration and then using the following theorem:

Theorem: If $\lim\limits_{x \to a} \ln (f(x)) = L,$ then $\lim\limits_{x \to a} f(x) = e^L.$

Q2. Find $\lim\limits_{x \to 0} x^{\sin x}.$ _____

Answer: If $y = x^{\sin x}$, then $\ln y = \ln x^{\sin x} = \sin x \ln x = \dfrac{\ln x}{\csc x}$. Hence $\lim\limits_{x \to 0} \ln y = \lim\limits_{x \to 0} \dfrac{\ln x}{\csc x} =$

$\lim\limits_{x \to 0}\left(\dfrac{\frac{1}{x}}{-\csc x \cot x}\right) = \lim\limits_{x \to 0}\left[\dfrac{\sin x}{x}(-\tan x)\right] = 1 \times 0 = 0$. Therefore, using the theorem above, $\lim\limits_{x \to 0} x^{\sin x} = e^0 = 1$.

Authors' Note: Remember that $\lim\limits_{x \to 0} \dfrac{\sin x}{x} = 1$. This "pops up" quite frequently in calculus.

Problem Set 3.11A

Calculate the indicated limits.

1. $\lim\limits_{x \to \pi/2} \left[(1 - \sin x)\tan x\right]$

2. $\lim\limits_{x \to 0} x \ln x$

3. $\lim\limits_{x \to \pi/2} \left[\cot x(1 + \sec x)\right]$

4. $\lim\limits_{x \to 0} \dfrac{3e^x}{x - 2}$

5. $\lim\limits_{x \to 1} x^{1/(1-x)}$

6. $\lim\limits_{x \to 0} \left(\dfrac{1}{x}\right)^{\sin x}$

7. $\lim\limits_{x \to \pi/2} (\sin x)^{\tan x}$

8. $\lim\limits_{x \to 2} (x - 1)^{1/(2-x)}$

Solutions and Comments for Problem Set 3.11A

1. $\lim\limits_{x \to \pi/2} \dfrac{-\cos x}{-\csc^2 x} = \lim\limits_{x \to \pi/2} \left[\cos x \sin^2 x\right] = 0 \times 1 = 0$.

Write $(1 - \sin x)\tan x$ as $\dfrac{1 - \sin x}{\cot x}$.

2. $\lim\limits_{x \to 0} \dfrac{1/x}{-1/x^2} = \lim\limits_{x \to 0} (-x) = 0$.

Write $x \ln x$ as $\dfrac{\ln x}{\frac{1}{x}}$.

3. Write $\cot x(1 + \sec x)$ as $\dfrac{1 + \sec x}{\tan x}$.

$\lim\limits_{x \to \pi/2} \dfrac{\sec x \tan x}{\sec^2 x} = \lim\limits_{x \to \pi/2} \sin x = 1$.

4. $\dfrac{3e^0}{0 - 2} = -\dfrac{3}{2}$.

This was just to see if you were awake. You didn't try to apply L'Hopital's Rule, did you???

5. $\lim\limits_{x \to 1} \ln x^{1/(1-x)} = \lim\limits_{x \to 1} \dfrac{\ln x}{1 - x} = \lim\limits_{x \to 1} \dfrac{\frac{1}{x}}{-1} = -1$;

hence $\lim\limits_{x \to 1} x^{1/(1-x)} = e^{-1} = \dfrac{1}{e}$.

6. $\lim\limits_{x\to 0} \ln\left(\dfrac{1}{x}\right)^{\sin x} = \lim\limits_{x\to 0}\left[\sin x\left(\ln\dfrac{1}{x}\right)\right]$

$= \lim\limits_{x\to 0}\left[\sin x(-\ln x)\right] = \lim\limits_{x\to 0}\dfrac{-\ln x}{\csc x}$

$= \lim\limits_{x\to 0}\left[\dfrac{-1/x}{-\csc x\cot x}\right] = \lim\limits_{x\to 0}\left[\dfrac{\sin x}{x}\cdot\tan x\right]$

$= 1\times 0 = 0$; hence $\lim\limits_{x\to 0}\left(\dfrac{1}{x}\right)^{\sin x} = e^0 = 1.$

7. $\lim\limits_{x\to \pi/2}\left[\tan x(\ln\sin x)\right] = \lim\limits_{x\to \pi/2}\dfrac{\ln\sin x}{\cot x}$

$= \lim\limits_{x\to \pi/2}\left[\dfrac{\dfrac{1}{\sin x}(\cos x)}{-\csc^2 x}\right] = \lim\limits_{x\to \pi/2}(-\sin x\cos x)$

$= -1\times 0 = 0.$ Desired limit is $e^0 = 1.$

8. $\lim\limits_{x\to 2}\dfrac{\ln(x-1)}{2-x} = \lim\limits_{x\to 2}\left[\dfrac{\dfrac{1}{x-1}}{-1}\right] = -1$; hence

desired limit is $e^{-1} = \dfrac{1}{e}.$

Problems 5–8 involve use of the theorem in the reading portion of this section. In (6), note that $\ln\left(\dfrac{1}{x}\right) = \ln 1 - \ln x = -\ln x$. It is *extremely important* to remember that L'Hopital's Rule cannot be applied until the form is $\dfrac{0}{0}$ or $\dfrac{\infty}{\infty}$.

---

*Section 3.12: Derivatives of Vector Functions and Parametrically Defined Functions

As discussed in Section 1.1A, a vector function $(x, y) = (f_1(t), f_2(t))$ can be written in parametric form $x = f_1(t)$, $y = f_2(t)$.

The derivative of such a vector function can be written $\left(\dfrac{dx}{dt}, \dfrac{dy}{dt}\right) = (f_1'(t), f_2'(t)).$

Authors' Note: Other notations may be used to express vector functions and their derivatives. If $\vec{R}$ represents a vector with initial point at the origin and tip (x, y), and if $\vec{i}$ and $\vec{j}$ represent the unit vectors $(1, 0)$ and $(0, 1)$ respectively, then

$$\vec{R} = x\vec{i} + y\vec{j}, \quad \text{and}$$

$$\dfrac{d\vec{R}}{dt} = \dfrac{dx}{dt}\vec{i} + \dfrac{dy}{dt}\vec{j}.$$

Q1. Find the derivative of the vector function $(x, y) = \left(\dfrac{1}{t}, \cos t\right).$ _____

Answer: $\left(\dfrac{dx}{dt}, \dfrac{dy}{dt}\right) = \left(-\dfrac{1}{t^2}, -\sin t\right).$

It is possible to find $\dfrac{dy}{dx}$ from parametric equations by using $\dfrac{dy}{dx} = \dfrac{\dfrac{dy}{dt}}{\dfrac{dx}{dt}}.$

Q2. Find $\dfrac{dy}{dx}$, given $x = 2t - 3$, $y = 3t^2$. _____

Answer: $\dfrac{dy}{dx} = \dfrac{6t}{2} = 3t$.

Authors' Note: In Q2, since $t = \dfrac{x + 3}{2}$, we could write $\dfrac{dy}{dx} = 3\left(\dfrac{x + 3}{2}\right) = \dfrac{3x + 9}{2}$.

Problem Set 3.12

1. Differentiate the vector function $(x, y) = (t - t^2, t^2 - 1)$.

2. Given $(x, y) = (1 + \cos 2t, \sin 2t)$.
   (a) Define the equation in Cartesian form.
   (b) Find $\left(\dfrac{dx}{dt}, \dfrac{dy}{dt}\right)$.

(c) Find $\dfrac{dy}{dx}$ two ways (in terms of x and in terms of t). Show that the two forms are equivalent.

3. Find $\dfrac{d^2\vec{R}}{dt^2}$ if $\vec{R} = \left(\dfrac{1}{t}, \sin t\right)$.

4. Find $\dfrac{d^2}{dt^2}[e^t\vec{i} + e^t \cos t\vec{j}\,]$.

Solutions and Comments for Problem Set 3.12

1. $(1 - 2t, 2t)$.

2. (a) $\sin^2 2t = y^2$, $\cos^2 2t = (x - 1)^2 \Rightarrow$
   $(x - 1)^2 + y^2 = 1$ or $y^2 = 2x - x^2$.
   (b) $(-2 \sin 2t, 2 \cos 2t)$.
   (c) $\dfrac{dy}{dx} = \dfrac{\cos 2t}{-\sin 2t} = \dfrac{x - 1}{-y} = \dfrac{1 - x}{y}$.

3. $\dfrac{d\vec{R}}{dt} = (-t^{-2}, \cos t) \Rightarrow \dfrac{d^2\vec{R}}{dt^2} = (2t^{-3}, -\sin t)$.

4. The first derivative is $e^t\vec{i} + (e^t \cos t - e^t \sin t)\vec{j}$. The second derivative is $e^t\vec{i} - 2e^t \sin t\vec{j}$.

The vector equation represents a circle with center $(1, 0)$ and radius $= 1$. For (c), differentiating $y^2 = 2x - x^2$ implicitly yields $2y\dfrac{dy}{dx} = 2 - 2x \Rightarrow \dfrac{dy}{dx} = \dfrac{1 - x}{y}$.

For the second derivative, $\dfrac{d}{dt}(e^t \cos t - e^t \sin t) = e^t \cos t - e^t \sin t - e^t \sin t - e^t \cos t = -2e^t \sin t$.

Section 3.13: Multiple Choice Questions for Chapter 3

1. $\displaystyle\lim_{h \to 0} \dfrac{\sin(\frac{\pi}{2} + h) - \sin\frac{\pi}{2}}{h} =$
   (A) 0
   (B) 1

   (C) -1
   (D) $\pi/2$
   (E) undefined

2. If $g(x) = 2x + 5$, then $\dfrac{d}{dx}(g^{-1}(x)) =$

   (A) $\dfrac{x-5}{2}$

   (B) $1/2$

   (C) 2

   (D) $2x - 5$

   (E) 4

3. If $y = e^{\sin x}$, then $\dfrac{dy}{dx} =$

   (A) $e^{\sin x}$

   (B) $e^{\cos x}$

   (C) $e^x \sin x$

   (D) $e^{\sin x} \cos x$

   (E) $e^{1-\sin x}$

4. $\lim\limits_{x \to 1} \dfrac{\ln x}{x^4 - 1} =$

   (A) 0

   (B) 1

   (C) $1/2$

   (D) $1/4$

   (E) Limit does not exist.

5.

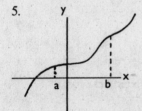
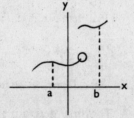
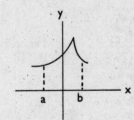
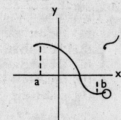

   The figures above show the graphs of four functions over the interval $[a, b]$. For *how many* of these functions could the Mean Value Theorem be applied on $[a, b]$?

   (A) 0

   (B) 1

   (C) 2

   (D) 3

   (E) 4

6. $\dfrac{d}{dx}(3x^2 + 4^3) =$

   (A) 0

   (B) $6x + 48$

   (C) $6x$

   (D) $6x + 4$

   (E) $3x^2$

7. If $xy + y = 3$, then $\dfrac{dy}{dx} =$

   (A) $\dfrac{-y}{1+x}$

   (B) 0

   (C) $3/y$

   (D) $\dfrac{3}{1+x}$

   (E) $-y$

8. Find $f'(x)$, given that $f(x) = \ln(\arcsin x)$.

   (A) $\dfrac{1}{\arcsin x}$

   (B) $\dfrac{\cos x}{\arcsin x}$

   (C) $\dfrac{\cos x}{\sqrt{1 - x^2}}$

   (D) $\dfrac{1}{\sqrt{1 - x^2}\,(\arcsin x)}$

   (E) $\dfrac{1}{1 + x^2}$

9. If $y = \sin u$, $u = v^2$, and $v = 4x$, then $\dfrac{dy}{dx} =$

   (A) $32x \cos 16x^2$

   (B) $8x \sin 16x^2$

   (C) $4x \sin 4x$

   (D) $32 \cos 8x^2$

   (E) None of the answers is correct.

10. If $g(x) = x^2 \cos x$, then $g'(x) =$
    (A) $2x \cos x$
    (B) $-x^2 \sin x$
    (C) $x^2 \sin x + 2x$
    (D) $2x \cos x - x^2 \sin x$
    (E) $2x \cos x + x^2 \sin x$

11. Given that $f(x) = |x - 3| + 2$, which one of the following statements is false?
    (A) f is continuous at $x = 3$
    (B) f is differentiable at $x = 3$
    (C) $f'(5) = 1$
    (D) $f'(0) = -1$
    (E) $f(2) = f(4)$

12. $\lim_{x \to 0} \dfrac{e^x + e^{-x} - 2}{3x^2} =$
    (A) 0
    (B) 1
    (C) 1/3
    (D) 1/6
    (E) limit does not exist

13. If $f(x) = e^{-2x}$, then $f^{(4)}(x) =$
    (A) $16e^{-x}$
    (B) $16e^{-2x}$
    (C) $-8e^{-2x}$
    (D) $8e^{-2x}$
    (E) $-16e^{-2x}$

14. *How many* of the following functions have the property that $f''(x) = -f(x)$?
    (i) $f(x) = \sin x$
    (ii) $f(x) = \cos x$
    (iii) $f(x) = e^{-x}$
    (iv) $f(x) = \ln x$
    (A) 0
    (B) 1
    (C) 2
    (D) 3
    (E) 4

15. $\dfrac{d}{dx}(\arccos(6x^2)) =$
    (A) $\dfrac{12x}{\sqrt{1 - 6x^2}}$
    (B) $\dfrac{x}{\sqrt{1 - 6x^2}}$
    (C) $\dfrac{1}{\sqrt{1 - 36x^4}}$
    (D) $\sin 12x$
    (E) $\dfrac{-12x}{\sqrt{1 - 36x^4}}$

16. If $y^2 - 3x = 7$, then $\dfrac{d^2y}{dx^2} =$
    (A) $\dfrac{-9}{4y^3}$
    (B) $\dfrac{3}{2y}$
    (C) 3
    (D) $\dfrac{-3}{y^3}$
    (E) $\dfrac{-6}{7y^3}$

17. If $y = \ln 4x + 4^x$, then $\dfrac{dy}{dx} =$
    (A) $\dfrac{4}{x} + x4^{x-1}$
    (B) $\dfrac{1}{x} + x4^{x-1}$
    (C) $\dfrac{1}{x} + \dfrac{1}{4}$
    (D) $\dfrac{1}{x} + 4\ln x$
    (E) $\dfrac{1}{x} + 4^x \ln 4$

18. Given that $f(x) = x^4 - 3$, find $c \in (0, 2)$ such that $\dfrac{f(2) - f(0)}{2 - 0} = f'(c)$.
    (A) 1
    (B) $\sqrt{2}$
    (C) $\sqrt{3}$
    (D) $\sqrt[3]{2}$
    (E) $\sqrt[4]{3}$

19. $\dfrac{d}{dx}(\sinh x + \cosh x) =$
    (A) $\dfrac{d}{dx}(\tanh x)$
    (B) e^x
    (C) $\dfrac{e^x + e^{-x}}{2}$
    (D) $\dfrac{e^x - e^{-x}}{2}$
    (E) None of the answers is correct.

20. $\lim_{h \to 0} \dfrac{3(x + h)^{37} - 3x^{37}}{h}$ is
    (A) the derivative of x^{38}
    (B) the derivative of $3x^{37}$
    (C) the derivative of x^{37}
    (D) equal to 3
    (E) undefined

21. If $y = u^3 + u$ and $u = x^2$, then $\dfrac{dy}{dx} =$
    (A) $x^3 + x$
    (B) $6x^5 + 2x$
    (C) $2x^4 + 2x^2$
    (D) $x^5 + x^2$
    (E) $5x^4 + 2x$

22. Given that f is a function, *how many* of the following statements are true?

(i) If f is continuous at $x = c$, then $f'(c)$ exists.

(ii) If $f'(c)$ exists, then f is continuous at $x = c$.

(iii) $\lim\limits_{x \to c} f(x) = f(c)$.

(iv) If f is continuous on (a, b), then f is continuous on $[a, b]$.

(A) 0

(B) 1

(C) 2

(D) 3

(E) 4

*23. $\lim\limits_{x \to 0} x^2 \ln x =$

(A) 0

(B) 1

(C) 2

(D) -2

(E) Limit does not exist.

*24.   Find $\dfrac{dy}{dx}$, given that $x = 8t - 12$ and $y = 2t^2 - 8t$.

(A) $\frac{1}{2}x - 1$

(B) $\dfrac{x - 4}{16}$

(C) $2x$

(D) $\dfrac{x}{16}$

(E) $\dfrac{x - 4}{2}$

*25. $\lim\limits_{x \to 1} x^{\frac{4}{1-x}} =$

(A) 0

(B) 1

(C) $1/e$

(D) $\dfrac{1}{e^4}$

(E) Limit does not exist.

Solutions and Comments for Multiple Choice Questions

1. **(A)**

> The expression is the derivative of $\sin x$ at the point where $x = \pi/2$.

> Indeterminate form $\dfrac{0}{0}$.

5. **(C)** The Mean Value Theorem could be applied to the graphs in the first and fourth figures.

> The third graph is not differentiable at the "peak."

2. **(B)** $g^{-1}(x) = \dfrac{x - 5}{2}$ and $\dfrac{d}{dx}\left(\dfrac{x - 5}{2}\right) = \dfrac{1}{2}$.

> If $y = 2x + 5$, then the derivative of the inverse is $\dfrac{dy}{dx} = \dfrac{1}{\dfrac{dy}{dx}} = \dfrac{1}{2}$.

6. **(C)**

> 4^3 is a constant and has a zero derivative.

3. **(D)**

> If $y = e^u$, then $\dfrac{dy}{dx} = e^u\left(\dfrac{du}{dx}\right)$.

7. **(A)** $x\dfrac{dy}{dx} + y + \dfrac{dy}{dx} = 0 \Rightarrow \dfrac{dy}{dx}(x + 1) = -y$

$\Rightarrow \dfrac{dy}{dx} = \dfrac{-y}{x + 1}$.

4. **(D)** Using L'Hopital's Rule, the existing limit is equal to $\lim\limits_{x \to 1}\left(\dfrac{1/x}{4x^3}\right) = \lim\limits_{x \to 1}\left(\dfrac{1}{4x^4}\right) = \dfrac{1}{4}$.

> A common error is to forget that the product rule for derivatives must be applied to the term xy.

8. **(D)** $f'(x) = \dfrac{1}{\arcsin x} \dfrac{d}{dx}(\arcsin x)$

$= \dfrac{1}{(\arcsin x)} \times \dfrac{1}{\sqrt{1-x^2}}.$

> If $y = \ln u$, then $\dfrac{dy}{dx} = \dfrac{1}{u} \times \dfrac{du}{dx}.$

9. **(A)** Using the chain rule, $\dfrac{dy}{dx} = \dfrac{dy}{du} \times \dfrac{du}{dv} \times \dfrac{dv}{dx}$

$= (\cos u)(2v)^4$, where $v = 4x$ and $u = v^2$.

10. **(D)** Use the product rule for derivatives.

11. **(B)** The graph of $y = f(x)$ follows.

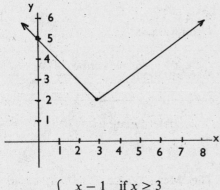

$f(x) = \begin{cases} x - 1 & \text{if } x \geq 3 \\ -x + 5 & \text{if } x < 3 \end{cases}$

12. **(C)** Given limit is equal to

$\lim\limits_{x \to 0} \dfrac{e^x - e^{-x}}{6x} = \lim\limits_{x \to 0} \dfrac{e^x + e^{-x}}{6} = \dfrac{2}{6} = \dfrac{1}{3}.$

> A double application of L'Hopital's Rule.

13. **(B)** $f'(x) = -2e^{-2x}$; $f''(x) = 4e^{-2x}$; $f'''(x) = -8e^{-2x}$; $f^{(4)}(x) = 16e^{-2x}$.

> The notation $f^{(4)}(x)$ is more convenient than $f''''(x)$.

14. **(C)** (i) and (ii) have the desired property.

> If $f(x) = e^{-x}$, then $f''(x) = f(x)$.

15. **(E)** $\cos y = 6x^2 \Rightarrow (-\sin y)\dfrac{dy}{dx} = 12x$

$\Rightarrow \dfrac{dy}{dx} = \dfrac{12x}{-\sin y}.$

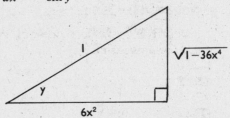

> Apply formula for $\dfrac{d}{dx}(\arccos u)$ in Section 3.6 as an alternative approach.

16. **(A)** $2y\dfrac{dy}{dx} - 3 = 0 \Rightarrow \dfrac{dy}{dx} = \dfrac{3}{2y} \Rightarrow \dfrac{d^2y}{dx^2}$

$= \dfrac{2y(0) - 3\left(2\dfrac{dy}{dx}\right)}{(2y)^2} = \dfrac{-6\left(\dfrac{3}{2y}\right)}{4y^2} = -\dfrac{9}{4y^3}.$

> You could also get a second derivative by starting with $\dfrac{dy}{dx} = 3(2y)^{-1}$, which yields $\dfrac{d^2y}{dx^2} = 3[(-1)(2y)^{-2}]\left(2\dfrac{dy}{dx}\right).$

17. **(E)** Note that derivative of 4^x is not $x4^{x-1}$.

18. **(D)** $\dfrac{f(2) - f(0)}{2} = \dfrac{13 - (-3)}{2} = 4x^3$

$\Rightarrow 8 = 4x^3 \Rightarrow x^3 = 2 \Rightarrow x = \sqrt[3]{2}.$

> Mean Value Theorem.

19. **(B)**

> Don't make this hard. $\sinh x + \cosh x = e^x$ and $\dfrac{d}{dx}(e^x) = e^x$.

20. **(B)** The expressed limit is equal to $111x^{36}$.

21. **(B)** $\dfrac{dy}{dx} = \dfrac{dy}{du} \times \dfrac{du}{dx} = (3u^2 + 1)(2x)$
$= (3x^4 + 1)(2x)$.

> You could also write $y = (x^2)^3 + x^2 = x^6 + x^2$ and then find $\dfrac{dy}{dx}$.

22. **(B)**

> The only correct statement is (ii). Differentiability implies continuity, but not vice versa. Statement (iii) implies continuity, which is not part of the stated premise. There are many examples to demonstrate that (iv) is not always true. For instance, consider the greatest integer function. It is continuous on $(0, 1)$, but not on $[0, 1]$.

23. **(A)** Write $x^2 \ln x$ as $\dfrac{\ln x}{\dfrac{1}{x^2}}$ and then use L'Hopital's Rule (Section 3.11A).

24. **(B)** $\dfrac{dy}{dx} = \dfrac{\dfrac{dy}{dt}}{\dfrac{dx}{dt}} = \dfrac{4t - 8}{8} = \dfrac{1}{2}t - 1$
$= \dfrac{1}{2}\left(\dfrac{x + 12}{8}\right) - 1 = \dfrac{x - 4}{16}$.

25. **(D)** $\ln y = \dfrac{4}{1 - x}\ln x = \dfrac{4(\ln x)}{1 - x}$. Hence
$\lim_{x \to 1} \ln y = \lim_{x \to 1} \dfrac{4(\ln x)}{1 - x} = \lim_{x \to 1}\left(\dfrac{4/x}{-1}\right)$
$= \lim_{x \to 1}\left(-\dfrac{4}{x}\right) = -4$. Hence, the desired limit
is $e^{-4} = \dfrac{1}{e^4}$.

> Use the theorem in Section 3.11A.

Chapter 4

APPLICATIONS OF DERIVATIVES

Section 4.1: Tangent and Normal Lines to a Curve and Linear Approximations

If f is a function that is differentiable at $x = a$, then $f'(a)$ is the slope of the tangent line to $y = f(x)$ at $(a, f(a))$. The equation of the tangent line is

$$y - f(a) = f'(a)(x - a).$$

Q1. Given $f(x) = x^3 + 6x + 4$. Find:

(a) The slope of the tangent line at the point where $x = 1$. _____

(b) The equation of the tangent line at $x = 1$. _____

Answers: (a) $f'(x) = 3x^2 + 6$; $f'(1) = 3 \times 1^2 + 6 = 9$.

       (b) $f(1) = 11$; hence equation of tangent line at $(1, 11)$ is $y - 11 = 9(x - 1)$, or $y = 9x + 2$.

• Authors' Note: Remember that a function can be continuous at $x = a$ and yet not be differentiable there. That is, continuity does not guarantee the existence of a unique tangent line. (See figure.)

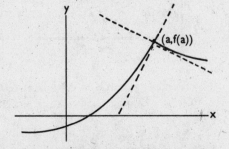

The *normal* to a graph at a point P is the line through P that is perpendicular to the tangent line at P. In geometry, one learns that two lines, neither of which is vertical, are perpendicular if and only if the product of their slopes is -1. Hence, it would follow that *(next page)*

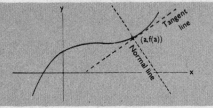

99

If f is a function that is differentiable at $x = a$, and if $f'(a) \neq 0$, then the equation of the normal line at $(a, f(a))$ is $y - f(a) = -\dfrac{1}{f'(a)}(x - a)$.

Q2. Find the equation of the normal line to $xy = 10$ at the point where $x = 2$. _____

Answer: If $x = 2$, then $y = 5$. Differentiating implicitly, $x\dfrac{dy}{dx} + y = 0 \Rightarrow \dfrac{dy}{dx} = -\dfrac{y}{x}$. At $(2, 5)$,

$\dfrac{dy}{dx} = -\dfrac{5}{2}$. Hence the slope of the normal line is $\dfrac{2}{5}$ and the equation of the normal line is

$y - 5 = \dfrac{2}{5}(x - 2)$.

Referring to the figure on the right, the slope of the tangent lines PS at $(x, f(x))$ is $f'(x) = \dfrac{dy}{dx}$. (The figure should remind the reader that the notation $\dfrac{dy}{dx}$ is consistent with the fact that the slope of the tangent line is $\tan \angle SPR$.)

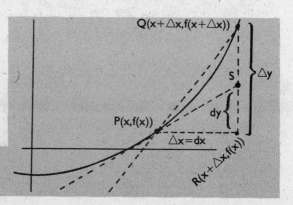

If one "solves" $\dfrac{dy}{dx} = f'(x)$, one obtains $dy = f'(x)\,dx$. dy is called the *differential* of y or $f(x)$.

Q3. What is the differential of y when $y = x^2 + \sin x$? _____

Answer: $\dfrac{dy}{dx} = 2x + \cos x$; hence $dy = (2x + \cos x)\,dx$.

The differential can be used to approximate a change in y for a "small" change in x. Referring to the figure above, as x changes from x to $(x + \Delta x)$ the actual change in y is

$$\Delta y = f(x + \Delta x) - f(x).$$

The approximate change in y is

$$dy = f'(x)\,dx.$$

Q4. If $y = x^2$, calculate:
    (a) the exact change in y as x increases from 2 to 2.01. _____
    (b) the approximate change, using the differential. _____

> Answers: (a) $(2.01)^2 - 2^2 = 4.0401 - 4 = 0.0401.$
> (b) $dy = 2x\,dx.$ If $x = 2$ and $dx = .01,$ then $dy = 2 \times 2(.01) = 0.04.$

Authors' Note: In general, if the change in x is relatively small, dy is a good approximation for the change in y. There are exceptions, of course. For instance, if $y = \sin\frac{1}{x}$, then dy cannot be counted on to give a good approximation for the change in y in any interval close to the origin.

Problem Set 4.1

In problems 1–5, find the equations of the tangent and normal lines at the point indicated.

1. $f(x) = x^4 - 3$ at $(2, 13)$

2. $y = \sin x$ at the point where $x = \frac{3\pi}{4}$

3. $y = x^2 - 7x + 4$ at the point where $x = 0$

4. $2xy + x = 0$ at the point where $x = 1$

5. $x^2 - xy + 6y - 6 = 0$ at $(1, -1)$

6. In each case, use differentials to find the approximate change in y.
   (a) $y = x^3 + 7x$ as x increases from 1 to 1.1
   (b) $y = 4x^2 - 8$ as x increases from 3 to 3.01

7. The edges of a cube are found to be 8.02 in. rather than the advertised dimension of 8 in. Use differentials to find
   (a) the approximate error in the surface area
   (b) the approximate error in the volume

8. Use differentials to calculate an approximation for $\sqrt{26}$.

Solutions and Comments for Problem Set 4.1

1. Tangent: $y - 13 = 32(x - 2)$;
   Normal: $y - 13 = -\dfrac{1}{32}(x - 2).$

2. Tangent: $y - \dfrac{\sqrt{2}}{2} = -\dfrac{\sqrt{2}}{2}\left(x - \dfrac{3\pi}{4}\right)$;
   Normal: $y - \dfrac{\sqrt{2}}{2} = \sqrt{2}\left(x - \dfrac{3\pi}{4}\right).$

3. Tangent: $y - 4 = -7x$; Normal: $y - 4 = \frac{1}{7}x.$

4. Tangent: $y = -\frac{1}{2}$; Normal: $x = 1.$

> When $x = 1$, $y = -\frac{1}{2}$; implicit differentiation yields $2x\dfrac{dy}{dx} + 2y + 1 = 0 \Rightarrow \dfrac{dy}{dx} = \dfrac{-(1 + 2y)}{2x}.$

5. Tangent: $y + 1 = -\frac{3}{5}(x - 1)$;
   Normal: $y + 1 = \frac{5}{3}(x - 1).$

> Implicit differentiation yields
> $$2x - \left(x\dfrac{dy}{dx} + y\right) + 6\dfrac{dy}{dx} = 0 \Rightarrow$$
> $$\dfrac{dy}{dx} = \dfrac{y - 2x}{6 - x}.$$

6. (a) $dy = (3x^2 + 7)\,dx$; when $x = 1$ and $dx = 0.1$, $dy = 10(0.1) = 1.$
   (b) $dy = 8x\,dx$; when $x = 3$, $dx = 0.01$, then $dy = 24(0.01) = 0.24.$

> (a) The actual change is 1.031.
> (b) The actual change is 0.2404.

7. (a) $S = 6x^2 \Rightarrow dS = 12x\,dx$; when $x = 8$ and $dx = 0.02$, $dS = 96(0.02) = 1.92$ (sq. in.).

(b) $V = x^3 \Rightarrow dV = 3x^2\,dx$; when $x = 8$ and $dx = 0.02$, $dV = 3(64)(0.02) = 3.84$ (cu. in.).

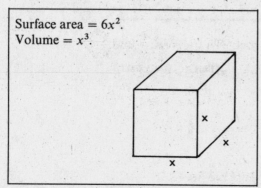

Surface area $= 6x^2$.
Volume $= x^3$.

8. $y = x^{1/2} \Rightarrow dy = \frac{1}{2}x^{-1/2}\,dx = \frac{dx}{2\sqrt{x}}$. When $x = 25$ and $dx = 1$, then $dy = \frac{1}{2\sqrt{25}} = \frac{1}{10} = 0.1$; hence an approximation for $\sqrt{26}$ is $\sqrt{25} + 0.1 = 5.1$.

> Basically, one just finds the approximate change in the function $y = \sqrt{x}$ as x increases from 25 to 26.

*Section 4.1A: Tangent Lines to Parametrically Defined Curves

If the parametric equations

$$x = f(t),$$
$$y = g(t)$$

are such that $f'(t)$ and $g'(t)$ exist, then for any t such that $f'(t) \neq 0$,

$$\frac{dy}{dx} = \frac{\dfrac{dy}{dt}}{\dfrac{dx}{dt}} = \frac{f'(t)}{g'(t)}.$$

Q1. If $x = t^2$ and $y = t^3 + 3$, find the equation of the tangent line when $t = 2$. _____

> Answer: If $t = 2$, then $x = 4$ and $y = 11$. $\dfrac{dy}{dx} = \dfrac{3t^2}{2t} = \dfrac{3t}{2}$; hence $\dfrac{dy}{dt} = 3$ when $t = 2$. The equation of the tangent line at $(4, 11)$ is $y - 11 = 3(x - 4)$.

Common error: If $x = t^2$ and $y = t^3 + 3$, then $\dfrac{dy}{dx} = 3t$ and $\dfrac{d^2 y}{dx^2} = 3$.

Authors' Note: The second derivative $\dfrac{d^2 y}{dx^2}$ is found by differentiating the first derivative $\dfrac{dy}{dx}$ with respect to x (not t). If $\dfrac{dy}{dx} = 3t$, then, using the chain rule, $\dfrac{d^2 y}{dx^2} = \dfrac{d}{dx}(3t) = \dfrac{d}{dt}(3t)\dfrac{dt}{dx} = \dfrac{\dfrac{d}{dt}(3t)}{\dfrac{dx}{dt}} = \dfrac{3}{2t}$.

Note that if $\dfrac{dy}{dx} = 3t$, then $\dfrac{d}{dt}\left(\dfrac{dy}{dx}\right) = 3$.

Problem Set 4.1A

In problems 1–4, find the equation of the tangent line at the point indicated.

1. $x = t + 1$, $y = t^2 - 4t$ at $t = 2$

2. $x = t^2$, $y = \dfrac{1}{t+2}$ at $t = 1$

3. $x = \sin t$, $y = \cos t$ at $t = \frac{\pi}{4}$

4. $x = t^4$, $y = t^2$ at $t = 1$.

In problems 5–7, find $\dfrac{dy}{dx}$ and $\dfrac{d^2y}{dx^2}$.

5. $x = t - t^2$, $y = t^2 - 1$

6. $x = 3\cos t$, $y = 4\sin t$

7. $x = 4t$, $y = 2 - 3t^2$

Solutions and Comments for Problem Set 4.1A

1. $\dfrac{dy}{dx} = \dfrac{2t - 4}{1} = 0$ when $t = 2$. Equation of tangent line is $y = -4$.

2. $\dfrac{dy}{dx} = \dfrac{\dfrac{-1}{(t+2)^2}}{2t} = \dfrac{-1}{2t(t+2)^2} = -\dfrac{1}{18}$ at $t = 1$. Equation of tangent line is $y - \frac{1}{3} = -\frac{1}{18}(x - 1)$.

3. $\dfrac{dy}{dx} = \dfrac{-\sin t}{\cos t} = -\tan t = -1$ at $t = \pi/4$. Equation of tangent line is $y - \dfrac{\sqrt{2}}{2} = -1\left(x - \dfrac{\sqrt{2}}{2}\right)$.

> If you recognize the unit circle $x^2 + y^2 = 1$, you merely have to find the line containing $\left(\dfrac{\sqrt{2}}{2}, \dfrac{\sqrt{2}}{2}\right)$ that is perpendicular to $y = x$.

4. $\dfrac{dy}{dx} = \dfrac{2t}{4t^3} = \dfrac{1}{2}$ at $t = 1$. Equation of tangent line is $y - 1 = \frac{1}{2}(x - 1)$.

5. $\dfrac{dy}{dx} = \dfrac{2t}{1 - 2t}$ and $\dfrac{d^2y}{dx^2} = \dfrac{\dfrac{d}{dt}\left(\dfrac{2t}{1 - 2t}\right)}{\dfrac{dx}{dt}} = \dfrac{2}{(1 - 2t)^3}$.

6. $\dfrac{dy}{dx} = \dfrac{4\cos t}{-3\sin t} = -\dfrac{4}{3}\cot(t)$ and $\dfrac{d^2y}{dx^2} = \dfrac{\dfrac{d}{dt}\left(-\dfrac{4}{3}\cot t\right)}{\dfrac{dx}{dt}} = \dfrac{-\dfrac{4}{3}(-\csc^2 t)}{-3\sin t} = -\dfrac{4}{9}\csc^3 t$.

7. $\dfrac{dy}{dx} = \dfrac{-6t}{4} = \dfrac{-3t}{2}$ and $\dfrac{d^2y}{dx^2} = \dfrac{\dfrac{d}{dt}\left(-\dfrac{3t}{2}\right)}{\dfrac{dx}{dt}}$

$= \dfrac{-\frac{3}{2}}{4} = -\dfrac{3}{8}$.

> If you are alert, it should not be surprising that $\dfrac{d^2y}{dx^2}$ turns out to be a constant. The parametric equations represent a parabola of the form $y = ax^2 + bx + c$, and y'' is a constant for any such parabola.

Section 4.2: Use of Derivatives in Curve Sketching (maximum, minimum, inflection points, concavity)

Increasing and Decreasing Functions

Given a function f such that $f'(x)$ exists at $x = x_1$:

(1) If $f'(x_1) > 0$, then $f(x)$ is *increasing* at $x = x_1$.

(2) If $f'(x_1) < 0$, then $f(x)$ is *decreasing* at $x = x_1$.

(3) If $f'(x_1) = 0$, then $f(x)$ is *stationary* at $x = x_1$ and the tangent line to $y = f(x)$ at $x = x_1$ is horizontal.

Authors' Note: If $f(x)$ is increasing on (a, b) and if x_1, $x_2 \in (a, b)$ such that $x_1 < x_2$, then $f(x_1) < f(x_2)$. The tangent line at any point in (a, b) would "rise" from lower left to upper right.

If $f(x)$ is decreasing on (b, c) and x_3, $x_4 \in (b, c)$ such that $x_3 < x_4$, then $f(x_3) > f(x_4)$ and the tangent lines will "fall" from upper left to lower right. In the figure, $f(x)$ is stationary at $x = b$.

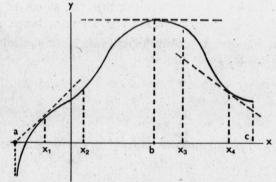

Q1. Given $f(x) = x^2 - 3x - 1$. Where is:

(a) $f(x)$ increasing? _____

(b) $f(x)$ decreasing? _____

(c) $f(x)$ stationary? _____

(d) Sketch the graph of f.

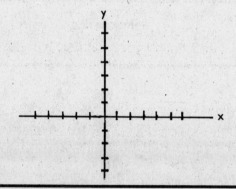

Answers: $f'(x) = 2x - 3$.

(a) $f(x)$ is increasing when $2x - 3 > 0$; that is, when $x \in (3/2, \infty)$.

(b) $f(x)$ is decreasing when $2x - 3 < 0$; that is, when $x \in (-\infty, 3/2)$.

(c) $f(x)$ is stationary when $2x - 3 = 0$; that is, when $x = 3/2$.

(d) The graph of f is on the right.

Concavity

If $f''(x) > 0$ for all $x \in (a, b)$, then the graph of $y = f(x)$ is *concave upward* on (a, b).

If $f''(x) < 0$ for all $x \in (a, b)$, then the graph of $y = f(x)$ is *concave downward* on (a, b).

Authors' Note: If $f''(x) > 0$, then $f'(x)$ is an increasing function. Geometrically, this means that the slopes of the tangent lines to $y = f(x)$ are increasing as x increases. Hence, one should expect a concave upward situation.

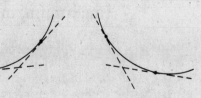

If $f''(x) < 0$, then the slopes of the tangent lines decrease as x increases; hence, a concave downward situation.

Q2. Find all intervals where $f(x) = 2x^3 - x^2 + 4x - 12$ is: (a) concave upward _____

(b) concave downward _____

> Answers: $f'(x) = 6x^2 - 2x + 4$ and $f''(x) = 12x - 2$.
> (a) $f''(x) > 0$ when $x > 1/6$; f is concave upward $(1/6, \infty)$.
> (b) $f''(x) < 0$ when $x < 1/6$; f is concave downward on $(-\infty, 1/6)$.

Relative Maximum and Minimum (Second Derivative Test)

If $f'(x_1) = 0$ and $f''(x_1) > 0$, then f has a *relative minimum* at $x = x_1$.

If $f'(x_1) = 0$ and $f''(x_1) < 0$, then f has a *relative maximum* at $x = x_1$.

Common error: Thinking that $f'(x_1) = 0$ necessitates a relative maximum or a relative minimum.

Authors' Note: If $f'(x_1) = 0$ and $f''(x_1) = 0$, then the second derivative test "fails" and other means must be used to check for a possible maximum or minimum at $x = x_1$. The function $y = x^3$ has the property that $y' = y'' = 0$ when $x = 0$; however, $x = 0$ is neither a relative maximum nor a relative minimum.

Q3. Find all relative maximum and minimum points for $f(x) = \dfrac{x^3}{3} - \dfrac{3x^2}{2} - 4x + 7$ and sketch the graph.

_____

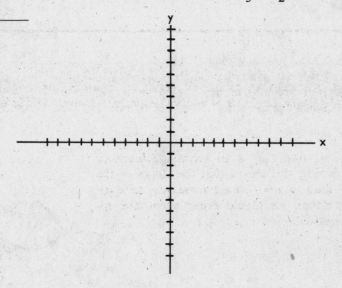

Answer: $f'(x) = x^2 - 3x - 4 = (x + 1)(x - 4)$.

$f'(x) = 0$ when $x = -1$, $x = 4$.

$f''(x) = 2x - 3$.

$f''(-1) = -5 < 0 \Rightarrow (-1, f(-1)) = (-1, 9\frac{1}{6})$ is a relative max.

$f''(4) = 8 > 0 \Rightarrow (4, f(4)) = (4, -11\frac{2}{3})$ is a relative min.

The graph is shown on the right.

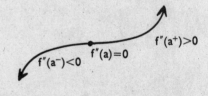

$(-1, 9\frac{1}{6})$ $(0, 7)$ $(4, -11\frac{2}{3})$

Common error: (Referring to Q3) ... A relative maximum value for $f(x)$ is -1.

Authors' Note: The relative max. occurs at $x = -1$. The actual value of the relative max. is $9\frac{1}{6}$. The point $(-1, 9\frac{1}{6})$ is a relative max. point. Moral of story: Watch use of terminology.

Inflection Point

A point where the direction of curvature changes is called a *point of inflection*. In Q3, an inflection point existed at $x = 3/2$. Given a function f, a point of inflection *may* occur when $f''(x) = 0$. In general, if $f''(a) = 0$ and if $f''(x)$ changes sign as x increases through $x = a$, then $(a, f(a))$ is a point of inflection.

Authors' Note: Some additional notation may be helpful here. If one lets a^- represent a number "slightly to the left" of a and a^+ represent a number "slightly to the right" of a, then one can write

$\left. \begin{array}{l} f''(a) = 0, f''(a^-) < 0, f''(a^+) > 0 \\ \qquad\qquad \text{or} \\ f''(a) = 0, f''(a^-) > 0, f''(a^+) < 0 \end{array} \right\} \Rightarrow \begin{array}{l} (a, f(a)) \text{ is a} \\ \text{point of} \\ \text{inflection.} \end{array}$

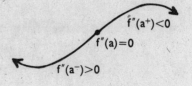

$f''(a^-) < 0$ $f''(a) = 0$ $f''(a^+) > 0$

$f''(a^+) < 0$ $f''(a) = 0$ $f''(a^-) > 0$

Note carefully: $f''(a) = 0$ does not by itself necessitate a point of inflection at $x = a$. For instance, if $f(x) = x^4$, then $f''(0) = 0$ but $(0, 0)$ is not a point of inflection.

Q4. Find all points of inflection for $f(x) = \sin 2x$ on $(0, 2\pi)$ and sketch the graph. _____

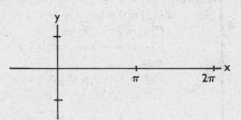

Answer: $f'(x) = 2\cos 2x$; $f''(x) = -4\sin 2x$. $f''(x) = 0$ when $2x = n\pi$ (n an integer). Hence $x = \dfrac{n\pi}{2}$. On $(0, 2\pi)$, one obtains

$x = \pi/2$, $x = \pi$ and $x = 3\pi/2$.

$f''\left(\dfrac{\pi^-}{2}\right) < 0$ and $f''\left(\dfrac{\pi^+}{2}\right) > 0 \Rightarrow \left(\dfrac{\pi}{2}, 0\right)$ is an inflection point.

$f''(\pi^-) > 0$ and $f''(\pi^+) < 0 \Rightarrow (\pi, 0)$ is an inflection point.

$f''\left(\dfrac{3\pi^-}{2}\right) < 0$ and $f''\left(\dfrac{3\pi^+}{2}\right) > 0 \Rightarrow \left(\dfrac{3\pi}{2}, 0\right)$ is an inflection point.

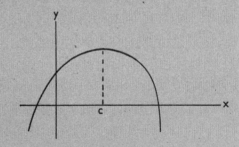

Absolute Maximum and Minimum Values

If $x_1 \in [a, b]$ and $f(x_1) \geq f(x)$ for all $x \in [a, b]$, then $f(x_1)$ is an *absolute maximum* on $[a, b]$.

If $x_1 \in [a, b]$ and $f(x_1) \leq f(x)$ for all $x \in [a, b]$, then $f(x_1)$ is an *absolute minimum* on $[a, b]$.

An absolute maximum (or minimum) can occur at $x = c$ when

(1) $f'(c) = 0$;

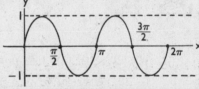

(2) c is in the domain of f, but $f'(c)$ is undefined;

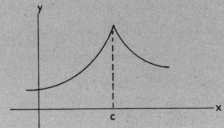

(next page)

or (3) c is an endpoint in the domain of f.

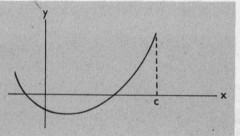

 Common error: Confusing relative max. (or min.) with absolute max. (or min.).

 Authors' Note: If $x = c$ produces a relative max. or min., then $f'(c) = 0$. However, if $x = c$ produces an absolute max. or min., it is not necessary that $f'(c) = 0$.

 Common error: If $f(x) = (x - 2)^{2/3} + 5$, then $f'(x) = \dfrac{2}{3(x - 2)^{1/3}}$. Since $f'(x) \neq 0$ for any real number, there is no maximum or minimum value for $f(x)$.

 Authors' Note: The number 2 is in the domain of f, but $f'(2)$ is not defined. However, $(x - 2)^{2/3} = \sqrt[3]{(x - 2)^2} \geq 0$ for any x and hence an absolute minimum for $f(x)$ will occur when $x = 2$. That is, $f(2) = 5$ is an absolute minimum.

General Summary:

Situation	What it means
$f'(a) > 0$	f increasing
$f'(a) < 0$	f decreasing
$f'(a) = 0$	Horizontal tangent
$f'(a) = 0, f'(a^-) < 0, f'(a^+) > 0$	Relative minimum
$f'(a) = 0, f'(a^-) > 0, f'(a^+) < 0$	Relative maximum
$f'(a) = 0, f''(a) > 0$	Relative minimum
$f'(a) = 0, f''(a) < 0$	Relative maximum
$f'(a) = 0, f''(a) = 0$	(?) . . . Need to explore further
$f''(a) > 0$	Concave upward

$f''(a) < 0$	Concave downward
$f''(a) = 0$	(?) ... Need to explore further
$f''(a) = 0, f''(a^-) > 0, f''(a^+) < 0$	Point of inflection
$f''(a) = 0, f''(a^-) < 0, f''(a^+) > 0$	Point of inflection
$f(a)$ exists, $f'(a)$ does not exist	Possibly a vertical tangent; possibly an absolute max. or min.
a is endpoint of a closed interval	$f(a)$ is possibly an absolute max. or min.

Problem Set 4.2

1. Using the figure, complete the chart by indicating whether the functions are positive, negative, or zero at the indicated points.

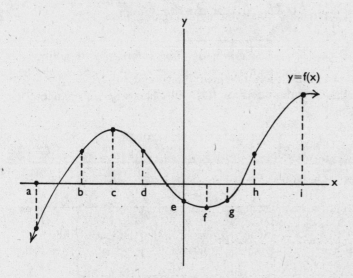

	$f(x)$	$f'(x)$	$f''(x)$
a			
b			
c			
d			
e			
f			
g			
h			
i			

2. Why must a cubic polynomial $f(x) = ax^3 + bx^2 + cx + d$, $a \neq 0$, always have a point of inflection?

3. For each function:
   (A) Identify all intervals on which the function is increasing.
   (B) Identify all intervals on which the function is decreasing.
   (C) Find all relative maximum and minimum values.
   (D) Identify all intervals on which the function is concave upward.
   (E) Identify all intervals on which the function is concave downward.
   (F) Find all points of inflection.
   (G) Find all absolute maximum and minimum values.

   (a) $f(x) = x^2 - 8x + 3$
   (b) $f(x) = \dfrac{x^3}{3} + x^2 - 3x + 1$
   (c) $f(x) = x^3 + 3x^2 - 24x + 20$ on $[0, 5]$
   (d) $g(x) = \cos \frac{x}{2}$ on $[0, 2\pi]$
   (e) $f(x) = (x - 1)^4$

4. Sketch the graph, identifying all maximum, minimum values and points of inflection.
   (a) $g(x) = x^3 - 12x$
   (b) $f(x) = 2x - x^2$
   (c) $f(x) = -x^3 + 3x^2 - 3x + 1$
   (d) $f(x) = x^4 - 8x^2 + 5$
   (e) $f(x) = \dfrac{1}{(x-1)^2}$

5. Find all absolute maximum and minimum points and sketch the graph.
   (a) $f(x) = |x + 2|$
   (b) $f(x) = -|x| + 4$
   (c) $f(x) = x^{1/3} + 1$
   (d) $f(x) = 2x^{2/3} - 3$

Solutions and Comments for Problem Set 4.2

1.

	$f(x)$	$f'(x)$	$f''(x)$
a	$-$	$+$	$-$
b	$+$	$+$	$-$
c	$+$	0	$-$
d	$+$	$-$	0
e	$-$	0	$+$
g	$-$	$+$	$+$
h	$+$	$+$	0
i	$+$	$+$	$-$

$f(c)$ is a relative max., $f(e)$ is a relative min., and $(d, f(d))$ and $(h, f(h))$ are points of inflection.

2. $f'(x) = 3ax^2 + 2bx + c$; $f''(x) = 6ax + 2b$. There will be a point of inflection when $6ax + 2b = 0$; that is, when $x = \dfrac{-b}{3a}$.

The graph of $f''(x) = 6ax + 2b$ is a line with nonzero slope $6a$; it intersects the x-axis at $\left(\dfrac{-b}{3a}, 0\right)$. Hence $f''(x)$ changes sign as x increases through $\dfrac{-b}{3a}$.

3.

	A Increasing (interval)	B Decreasing (interval)	C Rel. max., Rel. min. (points)	D Concave up (interval)	E Concave down (interval)	F Inflection (point)	G Abs. max., Abs. min. (points)
(a) $f'(x) = 2x - 8$ $f''(x) = 2$	$(4, \infty)$	$(-\infty, 4)$	$(4, -13)$ min.	$(-\infty, \infty)$	$\cdots$	None	$(4, -13)$ min.
(b) $f'(x) =$ $x^2 + 2x - 3$ $f''(x) = 2x + 2$	$(-\infty, -3)$ $\cup (1, \infty)$	$(-3, 1)$	$(-3, 10)$ max. $(1, -2/3)$ min.	$(-1, \infty)$	$(-\infty, -1)$	$(-1, 4\frac{2}{3})$	None
(c) $f'(x) =$ $3x^2 + 6x - 24$ $f''(x) = 6x + 6$	$(2, 5)$	$(0, 2)$	$(2, -8)$ min.	$(0, 5)$	$\cdots$	None	$(5, 100)$ max.

NOTE: f defined over closed interval domain. You *must* check endpoints.

(d) $g'(x) =$ $-\frac{1}{2}\sin\frac{x}{2}$ $g''(x) =$ $-\frac{1}{4}\cos\frac{x}{2}$	$\cdots$	$(0, 2\pi)$	None* *Derivative not defined at endpoints of closed interval.	$(\pi, 2\pi)$	$(0, \pi)$	$(\pi, 0)$	$(0, 1)$ max. $(2\pi, -1)$ min.
(e) $f'(x) =$ $4(x - 1)^3$ $f''(x) =$ $12(x - 1)^2$	$(1, \infty)$	$(-\infty, 1)$	$(1, 0)$ *not* rel. min. since $f'(1) = 0$ and $f''(1) = 0.$	$(-\infty, 1) \cup$ $(1, \infty)$	$\cdots$	None	$(1, 0)$ min.

4. (a)

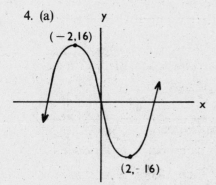

(b)

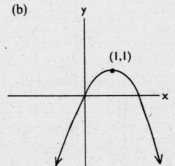

(c)

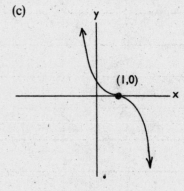

(d)

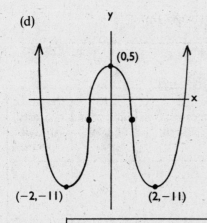

(e)

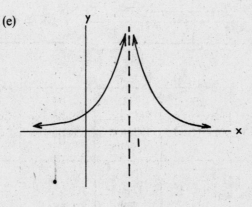

(a) $(-2, 16)$ rel. max., $(2, -16)$ rel. min., $(0, 0)$ inflection.

(b) $(1, 1)$ rel. and abs. max. No min., no points of inflection.

(c) No max. or min. $(1, 0)$ is point of inflection.

(d) $(0, 5)$ rel max. $(-2, -11)$ and $(2, -11)$ are rel. and abs. min. $\left(\dfrac{2\sqrt{3}}{3}, -\dfrac{35}{9}\right)$ and $\left(-\dfrac{2\sqrt{3}}{3}, -\dfrac{35}{9}\right)$ are points of inflection.

(e) No max., min., or inflection points. Note that (a) is an odd function and (d) and (e) are even functions. Recognizing this helps in the graphing process.

5. (a)

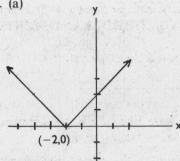

(b)

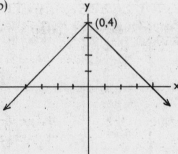

(c)

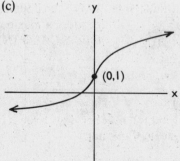

(d)

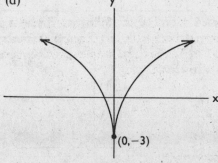

(a) $(-2, 0)$ abs. min.; (b) $(0, 4)$ abs. max.; (c) $(0, 1)$ inflection point; (d) $(0, -3)$ abs. min.

The derivative does not exist at any of the points above. In (c), the tangent line at $x = 0$ is vertical.

Section 4.3: Extreme Value Problems

Many practical-type problems requiring maximum and minimum values can be solved with derivatives.

One should keep in mind that extreme values can occur
  (1) at a zero of the first derivative;
  (2) at an endpoint of a closed interval;
  (3) at a point in the domain where the derivative is undefined.

Q1. Find two non-negative numbers whose sum is 20, such that the product of the numbers is a maximum.

Answer: Represent the numbers by x and $20 - x$. Then the product is $p(x) = x(20 - x) = 20x - x^2 \Rightarrow p'(x) = 20 - 2x$. Now $p'(x) = 0$ when $x = 10$ and $p''(10) = -2 < 0 \Rightarrow x = 10$ produces a maximum. The numbers are 10 and 10.

Authors' Note: In Q1, the domain for x is $[0, 20]$, a closed interval. The endpoints $x = 0$ and $x = 20$ do not produce a maximum value for $p(x)$. **DO NOT FORGET TO CONSIDER ENDPOINTS OF CLOSED INTERVALS.**

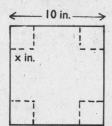

Problem Set 4.3

1. Find the number that exceeds its square by the greatest amount.

2. A square piece of tin has 10 in. on a side. An open box is formed by cutting out equal square pieces x in. on a side at the corners and bending upward the projecting portions which remain. Find the maximum volume that can be obtained.

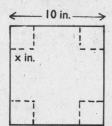

3. Find the least amount of lumber that will be needed to form an open box with a square base and a capacity of 32 cu. ft.

4. A manufacturer can ship a cargo of 200 tons at a profit of \$10 per ton. However, by waiting he can add 10 tons per week to the shipment, but the profit on the entire shipment will be reduced 20¢ per ton per week. When should the shipment be made to realize maximum profit?

5. Find the dimensions of the largest cylinder that can be cut from a solid sphere of radius a (where a is a constant).

6. The shaded region in the following figure is a rectangle with one vertex on the segment joining $(0, 1)$ and $(8, 0)$. Find the largest possible area for such a rectangle.

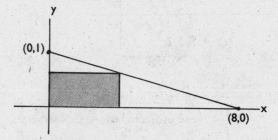

7. Find all maximum and minimum values for $f(x) = \sin x + \cos x$ on $[0, 2\pi]$.

8. Find all maximum and minimum values for $f(x) = x^2 e^x$.

Solutions and Comments for Problem Set 4.3

1. $y = x - x^2 \Rightarrow y' = 1 - 2x = 0$ when $x = \frac{1}{2}$.

> Use second derivative test to verify that $x = \frac{1}{2}$ produces a max.

2. $v(x) = x(10 - 2x)^2 = 4(x^3 - 10x^2 + 25x) \Rightarrow$ $v'(x) = 4(3x^2 - 20x + 25) = 4(3x - 5)(x - 5)$ $= 0$ when $x = 5/3$ or $x = 5$. Max. volume obtained when $x = 5/3$. The max. volume is $v(5/3) = \frac{2000}{27}$ (cu. in.).

> The domain for x is $(0, 5)$. Note that one of the zeros of the first derivative is not in the domain.

3. If x represents a side of the base and y represents the height, then $x^2 y = 32$ and the surface area is $s(x) = x^2 + 4xy = x^2 + 4x\left(\dfrac{32}{x^2}\right) = x^2 + 128x^{-1}$. Hence $s'(x) = 2x - 128x^{-2} = \dfrac{2(x^3 - 64)}{x^2} = 0$ when $x = 4$. Dimensions are 4 ft. by 4 ft. by 2 ft. and 48 sq. ft. is the minimum amount of lumber.

> Two variables are needed for this problem. However, one easily express one of the variables in terms of the other; that is, $x^2 y = 32 \Rightarrow y = \dfrac{32}{x^2}$.

4. If x represents the number of weeks to wait, then $p(x) = (200 + 10x)(10 - \frac{x}{5}) = 2000 + 60x - 2x^2 \Rightarrow p'(x) = 60 - 4x = 0$ when $x = 15$. Wait 15 weeks.

> This function and the one in problem #1 represent parabolas that open downward. Max. will occur at vertex of parabola.

6. $A = xy = x\left(-\frac{1}{8}x + 1\right) = -\frac{x^2}{8} + x \Rightarrow$ $A' = -\frac{x}{4} + 1 = 0$ when $x = 4$. Max. area $= 4 \times \frac{1}{2} = 2$ (sq. units).

> Equation of line joining $(0, 1)$ and $(8, 0)$ is $y = -\frac{1}{8}x + 1$.

5. Volume $= V = \pi r^2 h$. By the Pythagorean Theorem, $(2a)^2 = (2r)^2 + h^2 \Rightarrow r^2 = a^2 - \frac{1}{4}h^2$. Then $V = \pi(a^2 - \frac{1}{4}h^2)h$ and $V' = a^2\pi - \frac{3}{4}\pi h^2 = 0$ if $h = \dfrac{2a\sqrt{3}}{3}$; then $r = \dfrac{a\sqrt{6}}{3}$.

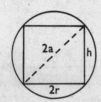

7. $f'(x) = \cos x - \sin x = 0$ when $\cos x = \sin x \Rightarrow \tan x = 1 \Rightarrow x = \pi/4$ or $x = 5\pi/4$ on $[0, 2\pi]$. $f(\pi/4) = \sqrt{2}$ is a max.; $f(5\pi/4) = -\sqrt{2}$ is a min.

> $f''(x) = -\sin x - \cos x$; $f''(\pi/4) < 0$; $f''(5\pi/4) > 0$.

8. $f'(x) = x^2 e^x + e^x(2x) = xe^x(x + 2) = 0$ when $x = 0$ or $x = -2$. Min. value $= f(0) = 0$; max value $= f(-2) = 4/e^2$.

> $x^2 e^x \geq 0$, so $(0, 0)$ is an obvious minimum point. It is important to note that the max. value at $x = -2$ is a relative maximum and not an absolute maximum over the set of real numbers. Note that $\lim_{x \to \infty} x^2 e^x = \infty$.

> $V' = \dfrac{dV}{dh}$; as in problem #3, one is faced with two variable quantities, r and h. Be sure to justify the assertion that the obtained values actually do produce a max.

Section 4.4: Velocity and Acceleration of a Particle Moving on a Line

If $y = s(t)$ represents a position function of a particle moving along a line, then
  (i) average velocity from t to $(t + \Delta t)$ is $\dfrac{\Delta s}{\Delta t} = \dfrac{s(t + \Delta t) - s(t)}{\Delta t}$, and
  (ii) instantaneous velocity at time t is $v(t) = s'(t)$, and
(iii) acceleration at time t is $a(t) = v'(t) = s''(t)$.

Q1. A particle moves along a line according to the position function $s(t) = t^4 - 6t^2 + 5$, where s represents the distance from the origin at time t.
  (a) Find the velocity function. _____
  (b) Find the acceleration function. _____
  (c) When is the particle stopped? _____
  (d) When is the particle at the origin? _____
  (e) Discuss the motion of the particle on the interval $[0, 3]$. _____

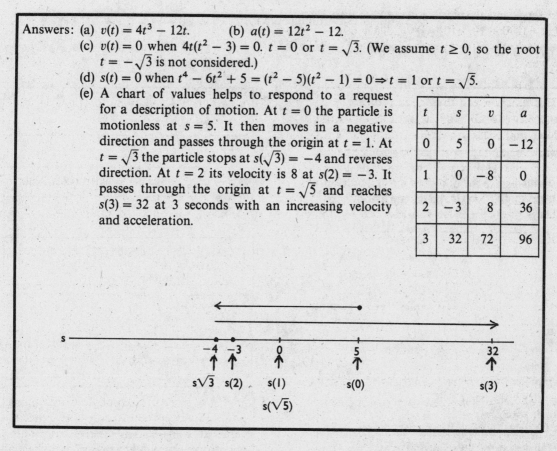

Answers: (a) $v(t) = 4t^3 - 12t$.   (b) $a(t) = 12t^2 - 12$.

(c) $v(t) = 0$ when $4t(t^2 - 3) = 0$. $t = 0$ or $t = \sqrt{3}$. (We assume $t \geq 0$, so the root $t = -\sqrt{3}$ is not considered.)

(d) $s(t) = 0$ when $t^4 - 6t^2 + 5 = (t^2 - 5)(t^2 - 1) = 0 \Rightarrow t = 1$ or $t = \sqrt{5}$.

(e) A chart of values helps to respond to a request for a description of motion. At $t = 0$ the particle is motionless at $s = 5$. It then moves in a negative direction and passes through the origin at $t = 1$. At $t = \sqrt{3}$ the particle stops at $s(\sqrt{3}) = -4$ and reverses direction. At $t = 2$ its velocity is 8 at $s(2) = -3$. It passes through the origin at $t = \sqrt{5}$ and reaches $s(3) = 32$ at 3 seconds with an increasing velocity and acceleration.

t	s	v	a
0	5	0	-12
1	0	-8	0
2	-3	8	36
3	32	72	96

Authors' Note: The description of motion can be as detailed as one wants to make it. It is generally useful to find when $s(t)$ and $v(t)$ are zero and to note when velocity is positive and when it is negative.

Problem Set 4.4

1. An object is hurled directly upward and its height above the ground is given by $s(t) = 64t - 16t^2$, where s is measured in feet and t in seconds.
   (a) When does the object reach its highest point?
   (b) What is the velocity when it hits the ground?
   (c) How far has it traveled when it hits the ground?

2. A particle moves along a line according to $s(t) = 2t^3 - 9t^2 + 12$, where $s(t)$ represents distance in feet from the origin after t seconds. Find
   (a) the velocity at $t = 2$
   (b) the acceleration at $t = 1$
   (c) when the particle is at rest
   (d) when the particle is moving in a positive direction

3. Motion along a line is described by $s(t) = (t^2 + 1)(t + 1)(t - 1)$, where s represents distance from the origin.
   (a) When is the object at the origin?
   (b) Find the velocity function.
   (c) When is the object at rest?
   (d) When is the velocity equal to 32?

4. A particle moves along a line according to $s(t) = t^4 - 4t^3 - 8t^2 + 2$, where $s(t)$ represents distance in feet from the origin after t seconds.
   (a) Find the average velocity on the interval $[0, 2]$.
   (b) At what time(s) is the particle at rest?

5. If $s(t) = 2t^3 - 6t^2 - 3t$ describes motion along a line, find the minimum and maximum velocities on the interval $[0, 3]$.

Solutions and Comments for Problem Set 4.4

1. (a) $v(t) = 64 - 32t$; $v(t) = 0$ when $t = 2$.
   (b) $v(4) = -64$ (ft./sec.).
   (c) $s(2) = 64$; $2 \times 64 = 128$ (ft.).

> Reaches highest point when velocity is 0. Note that $s(t)$ is a parabola opening downward; maximum point is the vertex. In (c), note that $s(4)$ is not the total distance traveled since the object changed direction during flight. $s(4)$ is the distance above the ground.

2. (a) $v(t) = 6t^2 - 18t$; $v(2) = -12$ (ft./sec.).
   (b) $a(t) = 12t - 18$ $a(1) = -6$ (ft./sec.^{2}).
   (c) $v(t) = 0$ at $t = 0$, $t = 3$.
   (d) $v(t) = 6t(t - 3) > 0$ when $t > 3$ (assuming $t \geq 0$).

3. (a) $s(t) = 0$ when $t = 1$ or $t = -1$ (If $t \geq 0$, ignore $t = -1$);
   (b) $s(t) = t^4 - 1 \Rightarrow v(t) = 4t^3$;
   (c) $v(t) = 0$ when $t = 0$.
   (d) $4t^3 = 32 \Rightarrow t = 2$.

> $t^2 + 1 > 0$ for all values of t. Writing $s(t)$ as $t^4 - 1$ makes it much easier to take the first derivative.

4. (a) $\dfrac{s(2) - s(0)}{2 - 0} = \dfrac{-46 - 2}{2} = -24$ (ft./sec.);
   (b) $v(t) = 4t^3 - 12t^2 - 16t = 4t(t^2 - 3t - 4)$
   $= 4t(t - 4)(t + 1) = 0$ when $t = -1, 0, 4$.
   Reject $t = -1$ if $t \geq 0$.

> *Common error :* Average velocity $= \dfrac{v(2) + v(0)}{2}$. Note that the average velocity over a time interval is *not* the average of the velocities at the endpoints of the interval.

5. $v(t) = 6t^2 - 12t - 3 \Rightarrow v'(t) = a(t) = 12t - 12$. $v'(t) = 0$ when $t = 1$; $v''(t) = 12$. $v''(1) > 0 \Rightarrow v(1)$ is a relative min. $v(0) = -3$, $v(1) = -9$ (min.), $v(3) = 15$ (max.).

> Remember ... if a function is continuous on a closed interval, a max. or min. can occur at the endpoints of the interval.

*Section 4.4A: Velocity and Acceleration Vectors for Motion in a Plane Curve

If a particle moves in the xy-plane, its motion at any instant is along the tangent line to the curve and its velocity is characterized by a magnitude and a direction. In the figure, the vectors $\overrightarrow{PV}$ and $\overrightarrow{OR}$ have the same magnitude (or length) and the same direction (designated by θ). The vectors are said to be *equal* and can be represented by the *position vector* (x, y).

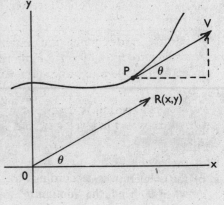

Authors' Note: Some individuals have difficulty relating to the concept of "equal" vectors. The vector (3, 1) represents an infinity of vectors constructed by starting at a point, moving 3 units to the right, 1 unit upward and then "drawing in the arrow." In a similar manner, the vector (−3, −1) is represented as shown on the right.

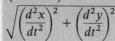

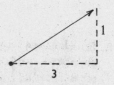

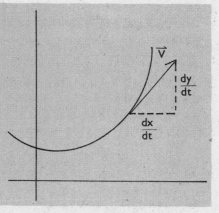

If the particle moves such that its coordinates at time t are $x = f(t)$, $y = g(t)$, then its *velocity* at time t is defined to be the vector

$$\vec{V} = (f'(t), g'(t)) = \left(\frac{dx}{dt}, \frac{dy}{dt}\right),$$

and its *speed* is $|\vec{V}| = \sqrt{\left(\frac{dx}{dt}\right)^2 + \left(\frac{dy}{dt}\right)^2}$.

In a similar manner, the acceleration is defined to be the vector $\left(\frac{d^2x}{dt^2}, \frac{d^2y}{dt^2}\right)$ and the magnitude of the acceleration is

$$\sqrt{\left(\frac{d^2x}{dt^2}\right)^2 + \left(\frac{d^2y}{dt^2}\right)^2}.$$

Q1. A particle moves along the curve $y = x^3$ in such a manner that its coordinates at time t are $x = t$, $y = t^3$. At $t = 1$ find its:

(a) velocity _____   (b) speed _____

(c) acceleration _____

> Answers: (a) $\vec{V} = (1, 3t^2)$. At $t = 1$, $\vec{V} = (1, 3)$.
> (b) Speed at $t = 1$ is $\sqrt{1^2 + 3^2} = \sqrt{10}$.
> (c) $\vec{A} = (0, 6t)$. At $t = 1$, $\vec{A} = (0, 6)$. The acceleration is a vector of magnitude 6 and direction 90°.

Problem Set 4.4A

1. A particle moves along the parabola $y = x^2$ according to $x = t$, $y = t^2$. Find the velocity, speed, and acceleration when (a) $t = 0$; (b) $t = 1$.

2. An object's motion is described by the vector $(2\sin t, 3\cos t)$. Find velocity and acceleration when $t = \pi/6$.

3. A particle moves according to $x = t - t^2$, $y = t^2 - 1$. Find the minimum speed of the particle.

4. A moving particle has position vector $(\ln(t - 1), t^2)$ for $t > 1$. Find the velocity and acceleration at $t = 2$.

5. An object's motion is described by the vector $(2\cos t + 3, 2\sin t + 3)$.
   (a) Show that the speed is constant.
   (b) What is the acceleration when $t = 0$?

6. A particle moves on the circle $x^2 + y^2 = 100$ in such a way that $x = t^2$, $t > 0$. Find the speed of the particle at $(8, -6)$.

Solutions and Comments for Problem Set 4.4A

1. $\vec{v} = (1, 2t)$ and $\vec{a} = (0, 2)$.
   (a) When $t = 0$, $\vec{v} = (1, 0)$, speed $= 1$ and $\vec{a} = (0, 2)$;
   (b) When $t = 1$, $\vec{v} = (1, 2)$, speed $= \sqrt{5}$ and $\vec{a} = (0, 2)$.

| Acceleration is constant. |
---

2. $\vec{v} = (2\cos t, -3\sin t)$ and
   $\vec{a} = (-2\sin t, -3\cos t)$. At $t = \pi/6$,
   $\vec{v} = \left(\dfrac{2\sqrt{3}}{2}, -\dfrac{3}{2}\right)$ and
   $\vec{a} = \left(-1, \dfrac{-3\sqrt{3}}{2}\right)$.

3. $\vec{v} = (1 - 2t, 2t)$ and $|v| = \sqrt{(1 - 2t)^2 + (2t)^2} = \sqrt{1 - 4t + 8t^2}$. $\dfrac{d|v|}{dt} = 0$ when $t = \dfrac{1}{4}$. At $t = \dfrac{1}{4}$,
   $|v| = \dfrac{\sqrt{2}}{2}$.

$\dfrac{dv}{dt} = \frac{1}{2}(1 - 4t + 8t^2)^{-1/2}(-4 + 16t)$. $\dfrac{dv}{dt} = 0$
when $-4 + 16t = 0$. You should, of course, test to see that $t = \frac{1}{4}$ does produce a minimum.

---

4. $\vec{v} = \left(\dfrac{1}{t - 1}, 2t\right)$ and $\vec{a} = \left(\dfrac{-1}{(t - 1)^2}, 2\right)$. At $t = 2$,
   $\vec{v} = (1, 4)$ and $\vec{a} = (-1, 2)$.

5. $\vec{v} = (-2\sin t, 2\cos t)$ and
   $\vec{a} = (-2\cos t, -2\sin t)$.
   $|v| = \sqrt{4\sin^2 t + 4\cos^2 t} = \sqrt{4} = 2$. At $t = 0$,
   $\vec{a} = (-2, 0)$.

| Never forget that $\sin^2 t + \cos^2 t = 1$ for any real number t. |
---

6. Differentiating with respect to t, $2x\dfrac{dx}{dt} + 2y\dfrac{dy}{dt} = 0 \Rightarrow x\dfrac{dx}{dt} + y\dfrac{dy}{dt} = 0$. $\dfrac{dx}{dt} = 2t$. When $x = 8$ and $y = -6$, we have $t = \sqrt{8} = 2\sqrt{2}$. Hence we have $8(4\sqrt{2}) + (-6)\dfrac{dy}{dt} = 0 \Rightarrow \dfrac{dy}{dt} = \dfrac{16\sqrt{2}}{2}$. Speed $= \sqrt{(4\sqrt{2})^2 + \left(\dfrac{16\sqrt{2}}{3}\right)^2} = \dfrac{20\sqrt{2}}{3}$.

| It's not unusual to find a problem similar to this one on standardized calculus tests. |

Section 4.5: Related Rates of Change

Example: A rectangular pool is 12 ft. long, 8 ft. wide, and 10 ft. deep. The pool is being filled with water. When the water is x ft. deep,

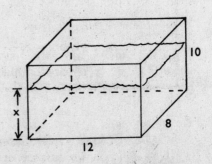

(a) The volume of water is $V(x) = 96x$.

(b) The rate of change of the volume of water is $\dfrac{dV}{dt}$.

(c) The rate of change of the depth is $\dfrac{dx}{dt}$.

(d) A formula relating $\dfrac{dV}{dt}$ and $\dfrac{dx}{dt}$ can be obtained by differentiating the equation in (a) with respect to t. That is, $\dfrac{dV}{dt} = 96\dfrac{dx}{dt}$.

We can now respond to questions such as this: If water is entering the pool at the rate of 3 cubic feet per second, how fast is the level of the water rising? To respond, we note that $3 = 96\dfrac{dx}{dt}$ and hence $\dfrac{dx}{dt} = \dfrac{1}{32}$ feet per second.

The following steps outline the method used to solve related rate problems.
(1) Draw and label a figure.
(2) State an equation which is valid at any time for the variables in the problem.
(3) Differentiate with respect to the appropriate variable (usually time).
(4) Solve this resulting equation for the variables desired, substituting into the equation the particular values specified in the problem or which have been calculated.

Q1. From the edge of a dock 12 feet above the surface of the water, a rowboat is being hauled in by a rope at the rate of 4 ft./sec. How fast is the length of rope changing when the boat is 5 ft. from the dock? (We will respond to this question in parts, using the figure on the right.)

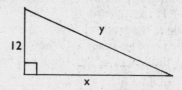

(a) State an equation involving x and y which is true at any time. _____

(b) Differentiate the equation with respect to the time variable t. _____

(c) Solve the equation in (b) for $\dfrac{dy}{dt}$. _____

(d) When $x = 5$, what are the values of y and $\dfrac{dx}{dt}$? _____

(e) Substitute the values from (d) into the equation in (c) to find $\dfrac{dy}{dt}$. _____

Answers: (a) $x^2 + 144 = y^2$. (b) $x\dfrac{dx}{dt} = y\dfrac{dy}{dt}$.

(c) $\dfrac{dy}{dt} = \dfrac{x}{y}\dfrac{dx}{dt}$. (d) When $x = 5$, we have $y = 13$ and $\dfrac{dx}{dt} = -4$.

(Since x is decreasing, $\dfrac{dx}{dt}$ is negative.)

(e) When $x = 5$, $\dfrac{dy}{dt} = \dfrac{5}{13}(-4) = -1\frac{7}{13}$ (ft./sec.).

(We say that the rope is shortening at the rate of $1\frac{7}{13}$ ft./sec.)

 Common error: Substituting specific values for the rates before differentiation, for instance, in (a) of Q1 writing $5^2 + 144 = y^2$. This would produce $\dfrac{dy}{dt} = 0$, which is not surprising since the equation erroneously states that y is a constant rather than a variable quantity.

Problem Set 4.5

1. A cube is compacted so that the volume decreases at a rate of 2 cu. meters per minute.
   (a) Find the rate of change of an edge of the cube when the volume is 27 cu. meters.
   (b) What is the rate of change of the surface of the cube at this point?

2. In the bottom of an hourglass, a conical pile of sand is formed at the rate of 12 in.^{3} per minute. The radius of the base of the pile is always equal to one-half its altitude. How fast is the altitude rising when it is 6 in. deep?

3. At night, a man 6 feet tall walks at a rate of 4 feet/sec. toward a streetlamp which is 10 feet above the ground. What is the rate of change of the tip of his shadow? What is the rate of change of the length of his shadow when he is 8 feet from the base of the light?

4. A ten-foot ladder leaning against a wall is pulled away from the wall at the rate of 2 feet/sec. How fast is the top of the ladder sliding down the wall when it is 8 feet above the ground?

5. A road and railroad track cross at right angles. A car going 60 mph crosses the track 10 min. before a high-speed train travelling at a rate of 90 mph. How fast is the distance between them changing 10 minutes after the train crosses the intersection?

6. In the following figure, the distance y is increasing at the rate of 10 units per minute. Find the rate of change of the angle θ when $y = 15$.

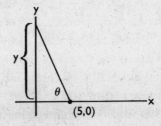

Solutions and Comments for Problem Set 4.5

1.

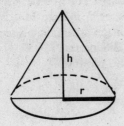

$V = \frac{1}{3}\pi r^2 h = \frac{1}{12}\pi h^3$ since $r = \frac{1}{2}h$. $\frac{dV}{dt} = \frac{\pi}{4}h^2\frac{dh}{dt}$.

When $\frac{dV}{dt} = 12$ and $h^2 = 36$, we have $\frac{dh}{dt} = \frac{4}{3\pi}$ (in./min.).

(a) $V = x^3 \Rightarrow \frac{dV}{dt} = 3x^2\frac{dx}{dt}$; when $\frac{dV}{dt} = 2$, $x = 3$, and $V = 27$, we have $\frac{dx}{dt} = \frac{2}{27}$ (m./sec.).

(b) $A = 6x^2 \Rightarrow \frac{dA}{dt} = 12x\frac{dx}{dt}$; when $x = 3$ and $\frac{dx}{dt} = \frac{2}{27}$, we have $\frac{dA}{dt} = \frac{8}{3}$ (sq. m./sec.).

2.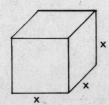

3.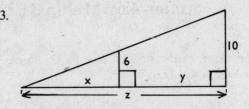

In the triangle, $\frac{6}{x} = \frac{10}{z} \Rightarrow z = \frac{5}{3}x$; also

$\frac{x+y}{x} = \frac{10}{6} \Rightarrow x = \frac{3}{2}y$. Hence

$\frac{dz}{dt} = \frac{5}{3}\frac{dx}{dt} = \frac{5}{3}(6) = 10$ (ft./sec.) and

$\frac{dx}{dt} = \frac{3}{2}\frac{dy}{dt} = \frac{3}{2}(4) = 6$ (ft./sec.).

An interesting aspect of this problem is that the answers are independent of the man's distance from the light.

$$2(15)(90) + 2(20)(60) = 2(25)\frac{dz}{dt} \Rightarrow \frac{dz}{dt}$$
$$= 102 \text{ (mph)}.$$

4.

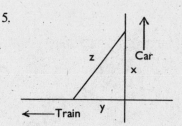

The car travels 10 miles before the train crosses the intersection and 10 miles after, so, at the desired instant, $x = 20$, $\frac{dx}{dt} = 60$, $y = 15$, $\frac{dy}{dt} = 90$, and $z = 25$.

$x^2 + y^2 = 100 \Rightarrow x\frac{dx}{dt} + y\frac{dy}{dt} = 0$. When $\frac{dx}{dt} = 2$, $y = 8$, and $x = 6$, we have $\frac{dy}{dt} = -\frac{3}{2}$ (ft./sec.).

6.

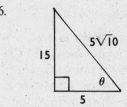

5.

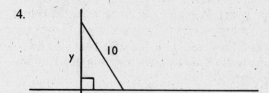

$x^2 + y^2 = z^2 \Rightarrow 2x\frac{dx}{dt} + 2y\frac{dy}{dt} = 2z\frac{dz}{dt}$. Hence

$\tan \theta = \frac{y}{5} \Rightarrow \sec^2 \theta \frac{d\theta}{dt} = \frac{1}{5}\frac{dy}{dt}$
$\Rightarrow \frac{d\theta}{dt} = \frac{1}{5}\cos^2 \theta \frac{dy}{dt}$. When $y = 15$,
$\cos \theta = \frac{1}{\sqrt{10}}$; hence, at this instant, $\frac{d\theta}{dt} = \frac{1}{5} \times$
$\frac{1}{10} \times 10 = \frac{1}{5}$ (radian/min.).

Section 4.6: Multiple Choice Questions for Chapter 4

1. The slope of the normal line to the curve $xy + 2y = 4$ at $(2, 1)$ is
   (A) 2
   (B) $-1/4$
   (C) -2
   (D) -1
   (E) 4

2. Given $f(x) = 3x^2 + x$, $x = 2$, and $dx = 0.002$, find dy.
   (A) 0.02
   (B) 0.026
   (C) 0.028

   (D) 0.014
   (E) 0.26

3. One leg of a right triangle begins to increase at the rate of 2 inches per minute while the other leg remains at 8 inches. In terms of in./min., how fast is the hypotenuse increasing when the first leg is 6 inches?
   (A) 2
   (B) 3
   (C) 6/5
   (D) 11/5
   (E) 4/3

4.

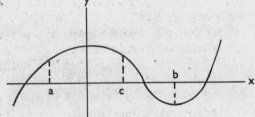

Given the function shown above, *how many* of the following statements are true?

    (i) $f'(b) = 0$
    (ii) $f''(a) < 0$
    (iii) $f''(c) < 0$
    (iv) $f''(b) > 0$

(A) 0
(B) 1
(C) 2
(D) 3
(E) 4

5. The equation of the tangent line to $y = x^2 - 6$ at the point where $x = -2$ is

(A) $y = -4x - 10$
(B) $y = -2x$
(C) $x + 4y = 10$
(D) $y = x + 6$
(E) $2y + x = -4$

6. The graph of $y = 2x^3 + 5x^2 - 6x + 7$ has a point of inflection at $x =$

(A) $-5/3$
(B) 0
(C) $-5/6$
(D) $5/2$
(E) -2

7. If $y = -x^2 + 4x + 25$, what is the maximum value for y?

(A) 25
(B) -16
(C) 28
(D) 29
(E) 18

8. A particle moves along a line according the position function $s(t) = t^3 + 4t - 3$. What is the velocity of the particle at $t = 2$?

(A) 13
(B) 14
(C) 16

(D) 24
(E) 30

9. If the graph of $y = ax^3 + 4x^2 + cx + d$ has a point of inflection at $(1, 0)$, then the value of a is

(A) 2
(B) $-4/3$
(C) $1/2$
(D) $2/3$
(E) $9/7$

10. What is the absolute minimum value of the function $g(x) = xe^x$?

(A) -1
(B) $-e$
(C) $-1/e$
(D) $-1/e^2$
(E) $-1/2$

11. If $f(x) = \cos 3x$, find all values in the interval $(0, \pi)$ for which $f^{(3)}(x) = 0$.

(A) $\pi/6$
(B) $\pi/4$
(C) $\pi/3, 2\pi/3$
(D) $\pi/6, 5\pi/6$
(E) $\pi/2$

12. The normal line to $y = 2 \sin x$ at $x = \pi/3$ intersects the x-axis at $(x_1, 0)$. What is the value of x_1?

(A) $1/2$
(B) $\dfrac{\pi + 6}{3}$
(C) $2\pi/3$
(D) $\dfrac{\pi + 3\sqrt{3}}{3}$
(E) $\pi/6$

13.

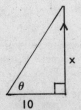

Consider the figure above where the distance x is increasing at the rate of 50 units per second. In radians per second, what is the rate of change of the angle θ when $x = 10$?

(A) 1
(B) 1.25
(C) 1.5
(D) 2
(E) 2.5

14. Given that $g(x) = -2x^3 - 6x^2 - 7$, what is the minimum value of $g(x)$ on the interval $[1, 2]$?
(A) -47
(B) -15
(C) -7
(D) 0
(E) 4

15. Given $f(x) = x^3 - 6x^2$. Over which one of the five intervals shown is f decreasing?
(A) $[-4, -2]$
(B) $[1, 3]$
(C) $[6, 8]$
(D) $[-1, 0]$
(E) $[4, 5]$

16. A function $y = f(x)$ has the properties $f'(a) = 0$ and $f''(a) = 0$. Which one of the following statements is true?
(A) The graph of $y = f(x)$ has a horizontal tangent at $(a, f(a))$.
(B) $(a, f(a))$ is a point of inflection.
(C) $(a, f(a))$ must be either a maximum or a minimum point.
(D) f may be discontinuous at $x = a$.
(E) None of the above is necessarily true.

17. Given a function $w(x) = -2x^2 + 16x + a$. What is the value of a if the absolute maximum value of the function is 39?
(A) 16
(B) 36
(C) 7
(D) 18
(E) 13

18. The motion of a particle along a line is given by the position function $s(t) = \dfrac{t^3}{3} - \dfrac{9t^2}{2} + 14t - 6$. At what time(s) in the interval $[0, 10]$ is the particle at rest?
(A) $t = 2, t = 7$
(B) $t = 3$
(C) $t = 4, t = 6$
(D) $t = 8$
(E) $t = 5/2, t = 6$

19. Given a function f, *how many* of the following statements are true?
(i) If $f''(a) < 0$, then the graph of $y = f(x)$ is concave upward at $x = a$.
(ii) If $f'(a)$ does not exist, then a is not in the domain of f.
(iii) If $f'(a) = 0$ and $f''(a) > 0$, then $f(a)$ is a relative maximum value.
(iv) If $f'(a) = 0$ and $f''(a) = 0$, then $f'''(a) = 0$.
(A) 0
(B) 1
(C) 2
(D) 3
(E) 4

20. The figure shown is a rectangle enclosed by a semicircle of radius 1 ft. What is the maximum area of such a rectangle?

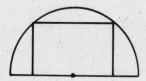

(A) 2 sq. ft.
(B) $1\frac{1}{2}$ sq. ft.
(C) 1.25 sq. ft.
(D) 1 sq. ft.
(E) 0.75 sq. ft.

21. If $g(x)$ is a second degree polynomial satisfying $g(0) = 3$, $g'(2) = 10$, and $g''(10) = 4$, then $g(x) =$
(A) $3x^2 + 3$
(B) $x^2 + 4x + 3$
(C) $2x^2 + 4x + 3$
(D) $2x^2 + 2x + 3$
(E) $x^2 + 6x + 3$

*22. If $x = 2t - 7$ and $y = 8t^2$, then $\dfrac{d^2y}{dx^2} =$
(A) $8t$
(B) $16t$
(C) 4
(D) 8
(E) $4t$

*23. If $x = 4t$ and $y = \sin t$, what is the equation of the tangent line at $t = \pi/2$?
(A) $y = x$

(B) $y - 1 = \frac{\pi}{2}(x - 1)$

(C) $y = x + \pi/2$

(D) $y = 1$

(E) $y = 4x - \pi/2$

*24. If $x^2 + 2y = 20$ and $\frac{dx}{dt} = -6$, find $\frac{dy}{dt}$ when

$x = 4$.

(A) 4

(B) -2

(C) -6

(D) 24

(E) 8

*25. A particle moves according to $x = t^2 + 4t$, $y = 1 - 6t$. What is the speed of the particle at $t = 1$?

(A) 8

(B) $6\sqrt{2}$

(C) $4\sqrt{3}$

(D) 12

(E) $3\sqrt{5}$.

Solutions and Comments for Multiple Choice Questions

1. **(E)** $x\frac{dy}{dx} + y + 2\frac{dy}{dx} = 0 \Rightarrow \frac{dy}{dx} = \frac{-y}{x + 2}$.

   At $(2, 1)$, $\frac{dy}{dx} = -\frac{1}{4} \Rightarrow$ Slope of normal line is 4.

   > Normal line is perpendicular to tangent line.

2. **(B)** $dy = (6x + 1)dx$. Substituting appropriate values yields $dy = 13(.002) = 0.026$.

3.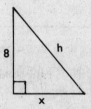

   (C) $8^2 + x^2 = h^2 \Rightarrow 2x\frac{dx}{dt} = 2h\frac{dh}{dt} \Rightarrow$

   $\frac{dh}{dt} = \frac{x}{h} \times \frac{dx}{dt}$. When $x = 6$, $h = 10$, and

   $\frac{dx}{dt} = 2$, we have $\frac{dh}{dt} = \frac{12}{10} = \frac{6}{5}$.

4. **(E)**

   > Make sure you can interpret the signs of the first and second derivatives.

5. **(A)** $\frac{dy}{dx} = 2x$. At $x = -2$, we have $\frac{dy}{dx} = -4$.
   Desired equation is $y + 2 = -4(x + 2)$.

6. **(C)** $y'' = 12x + 10$, which is equal to zero when $x = -5/6$.

   > A third degree polynomial has exactly one point of inflection.

7. **(D)** $y' = -2x + 4 = 0$ when $x = 2$. $-(2)^2 + 4(2) + 25 = 29$.

   > Function is a parabola opening downward. Max. value will occur at vertex.

8. **(C)** $v(t) = 3t^2 + 4$; $v(2) = 16$.

9. **(B)** $y'' - 6ax + 8 = 0$ when $x = \frac{-4}{3a}$. Setting

   $\frac{-4}{3a} = 1$ yields $a = -\frac{4}{3}$.

   > See comment for problem #6.

10. **(C)** $g'(x) = xe^x + e^x = e^x(x + 1) = 0$ when $x = -1$. $g(-1) = -1/e$.

    > Easy to establish that $g''(-1) > 0$.

11. **(C)** $f^{(3)}(x) = 27 \sin 3x = 0$ when $\sin 3x = 0$, or when $x = \frac{\pi}{3} \times n$, where n is an integer.

    > Note carefully that $(0, \pi)$ is an open interval.

12. **(D)** Equation of normal line is $y - \sqrt{3} = -1(x - \pi/3)$.

> Set $y = 0$ in equation of normal line and solve for x.

13. **(E)** $\tan\theta = \dfrac{x}{10} \Rightarrow \sec^2\theta\dfrac{d\theta}{dt} = \dfrac{1}{10}\dfrac{dx}{dt}$.
When $x = 10$, we have $\sec^2\theta = 2$.
Hence $2\dfrac{d\theta}{dt} = \dfrac{1}{10}(50) \Rightarrow \dfrac{d\theta}{dt} = 5/2$.

14. **(A)** $g(2) = -47$.

> Closed interval, hence you must check endpoints for max., min. values.

15. **(B)** $f'(x) = 3x^2 - 12x = 3x(x - 4) < 0$ if $0 < x < 4$.

> Note that $[1, 3] \subset (0, 4)$.

16. **(E)**

> Note that (A), (B), and (C) could be true, but they are not a necessary consequence of the premise.

17. **(C)** $w'(x) = -4x + 16 = 0$ if $x = 4$. $w(4) = 39 \Rightarrow a = 7$.

> Graph is a parabola opening downward, so max. value will occur at vertex.

18. **(A)** $v(t) = t^2 - 9t + 14 = (t - 2)(t - 7)$.

> Particle at rest when $s'(t) = v(t) = 0$.

19. **(A)**

> As an example of a function that has 0 in its domain, but $f'(0)$ is undefined, consider $f(x) = x^{1/2}$.

20.

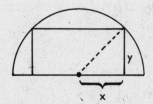

(D) $A = 2xy = 2x\sqrt{1 - x^2}$.
$\dfrac{dA}{dx} = (x^2 - x^4)^{-1/2}(2x - 4x^3) = 0$
when $2x(1 - 2x^2) = 0$. Solving, $x = 0$ (reject), $x = \sqrt{\frac{1}{2}}$. Max. area is $2\sqrt{\frac{1}{2}}\sqrt{1 - \frac{1}{2}} = 1$.

> For differentiating, it is helpful to write $2x\sqrt{1 - x^2}$ as $2(x^2 - x^4)^{1/2}$.

21. **(D)** Let $g(x) = ax^2 + bx + c$. Then $g'(x) = 2ax + b$ and $g''(x) = 2a$. $g(0) = 0 \Rightarrow c = 3$. $g'(2) = 10 \Rightarrow 4a + b = 10$. $g''(10) = 4 \Rightarrow a = 2$; hence $b = 2$.

22. **(C)** $\dfrac{dy}{dx} = \dfrac{16t}{2} = 8t$. $\dfrac{d^2y}{dx^2} = \dfrac{d}{dt}(8t)\dfrac{dt}{dx} = \dfrac{8}{\dfrac{dx}{dt}} = \dfrac{8}{2} = 4$.

> One should expect $\dfrac{d^2y}{dx^2}$ to be a constant since one can observe that if y is expressed in terms of x, the form would be
> $$y = ax^2 + bx + c.$$

23. **(D)** $\dfrac{dy}{dx} = \dfrac{\cos t}{4} = 0$ when $t = \pi/2$. Tangent line at $(2\pi, 1)$ is $y - 1 = 0(x - 2\pi)$.

24. **(D)** $2x\dfrac{dx}{dt} + 2\dfrac{dy}{dt} = 0 \Rightarrow \dfrac{dy}{dt} = -x\dfrac{dx}{dt}$.
When $x = 4$, $\dfrac{dy}{dt} = 24$.

25. **(B)** $\dfrac{dx}{dt} = 2t + 4$, $\dfrac{dy}{dt} = -6$. Hence $|v| = \sqrt{(2t + 4)^2 + (-6)^2}$. At $t = 1$, $|v| = 6\sqrt{2}$.

> The velocity vector $\vec{v} = (2t + 4, -6)$. At $t = 1$, $\vec{v} = (6, -6)$.

Chapter 5

THE INDEFINITE INTEGRAL: TECHNIQUES AND APPLICATIONS

Section 5.1: The Indefinite Integral and Basic Integration Formulas

If F and f are functions such that $F'(x) = f(x)$, then F is called an *antiderivative* or *indefinite integral* of f and is denoted by $\int f(x)\,dx$. The symbol $\int$ is called an *integral sign* and $f(x)$ is called the *integrand*. If $F(x)$ is an antiderivative of $f(x)$, then so is $F(x) + C$, where C is any constant, for $\frac{d}{dx}(F(x) + C) = F'(x) + \frac{d}{dx}(C) = f(x) + 0 = f(x)$. Hence, if $F'(x) = f(x)$, then $\int f(x)\,dx = F(x) + C$.

Basic integration formulas appear below. One can verify them by taking derivatives of the functions on the right-hand side of the equation. Those marked with an * are encountered quite frequently and should be well known by the reader.

*(1) $\int u^n\,du = \dfrac{u^{n+1}}{n+1} + C$ if $n \neq 1$

*(2) $\int \dfrac{du}{u} = \ln|u| + C$

*(3) $\int \sin u\,du = -\cos u + C$

*(4) $\int \cos u\,du = \sin u + C$

(5) $\int \tan u\,du = \ln|\sec u| + C$

(6) $\int \cot u\,du = \ln|\sin u| + C$

(7) $\int \sec u\,du = \ln|\sec u + \tan u| + C$

(8) $\int \csc u\,du = \ln|\csc u - \cot u| + C$

*(9) $\int \sec^2 u\,du = \tan u + C$

(10) $\int \csc^2 u\,du = -\cot u + C$

(11) $\displaystyle\int \sec u \tan u \, du = \sec u + C$

(12) $\displaystyle\int \csc u \cot u \, du = -\csc u + C$

*(13) $\displaystyle\int e^u \, du = e^u + C$

(14) $\displaystyle\int a^u \, du = \frac{a^u}{\ln a} + C \quad (a > 0, a \neq 1)$

*(15) $\displaystyle\int \frac{du}{u^2 + a^2} = \frac{1}{a}\tan^{-1}\left(\frac{u}{a}\right) + C$

(16) $\displaystyle\int \frac{du}{u^2 - a^2} = \frac{1}{2a}\ln\left|\frac{u - a}{u + a}\right| + C \quad (u > a)$

(17) $\displaystyle\int \frac{du}{a^2 - u^2} = \ln\left|\frac{a + u}{a - u}\right| + C \quad (u < a)$

*(18) $\displaystyle\int \frac{du}{\sqrt{a^2 - u^2}} = \sin^{-1}\left(\frac{u}{a}\right) + C$

The following are fundamental properties of the indefinite integral:

*(19) $\displaystyle\int kf(u) \, du = k\int f(u) \, du \quad$ for any constant k.

*(20) $\displaystyle\int [f(u) + g(u)] \, du = \int f(u) \, du + \int g(u) \, du.$

Authors' Note: $\displaystyle\int f(u)g(u) \, du$ is *not* equal to $\displaystyle\int f(u) \, du \times \int g(u) \, du.$

Q1. Evaluate $\displaystyle\int \left(x^3 + \sin x + \frac{1}{x} - \sec^2 x\right) dx.$ _____

> Answer: Using properties and formulas (20), (1), (3), (2), and (9), one obtains $\dfrac{x^4}{4} - \cos x + \ln|x| - \tan x + C.$

The process of evaluating an indefinite integral is called *integration*.

  Common error: Forgetting the constant of integration and writing $\dfrac{x^4}{4} - \cos x + \ln x - \tan x$ as the answer for Q1.

Authors' Note: $\displaystyle\int f(x) \, dx$ represents the "family" of functions whose members have the property that their derivatives are equal to $f(x)$. Members in this "family" can differ by a constant.

Q2. Find $f(x)$ if $f'(x) = 2x + 7$ and the graph of f contains the point $(1, 12)$. _____

> Answer: $\displaystyle\int (2x + 7) \, dx = x^2 + 7x + C; f(1) = 12 \Rightarrow 1^2 + 7\cdot 1 + C = 12 \Rightarrow C = 4.$
> Hence $f(x) = x^2 + 7x + 4.$

Problem Set 5.1

In problems 1–9, evaluate the indefinite integral.

1. $\int (x^4 + x^3 + x + 1)\,dx$

2. $\int \left(x^2 + \dfrac{1}{x^2} \right) dx$

3. $\int (w^2 + 1)^2 \, dw$

4. $\int (\sqrt{t} + 4)\,dt$

5. $\int \left(\dfrac{3 + x^2}{x^2} \right) dx$

6. $\int (x^{1/3} + x^{-5/3})\,dx$

7. $\int (\sin x + \cos x)\,dx$

8. $\int \left(e^z + \dfrac{1}{z} \right) dz$

9. $\int \sqrt[5]{x^2}\,dx$

In problems 10–13, find $f(x)$ using the given information.

10. $f'(x) = x^2 + 4x$ and f contains the point $(0, 14)$

11. $f'(x) = \cos x$ and $f(\frac{\pi}{2}) = 5$

12. $f'(x) = \frac{5}{x}$ and $f(e^2) = 7$

13. $f'(x) = \sec^2 x$ and $f(\frac{\pi}{4}) = -3$

Solutions and Comments for Problem Set 5.1

1. $\dfrac{x^5}{5} + \dfrac{x^4}{4} + \dfrac{x^2}{2} + x + C.$

2. $\dfrac{x^3}{3} - \dfrac{1}{x} + C.$

> Write $\int \left(x^2 + \dfrac{-1}{x^2} \right) dx$ as $\int (x^2 + x^{-2})\,dx.$
>
> Note how fractional exponents and negative exponents are very useful in the integration process. Remember that one can check the results of integration by taking a derivative.

3. $\dfrac{w^5}{5} + \dfrac{2w^3}{3} + w + C.$

4. $\frac{2}{3}t^{3/2} + 4t + C.$

5. $\int (3x^{-2} + 1)\,dx = -\dfrac{3}{x} + x + C.$

6. $\frac{3}{4}x^{4/3} - \frac{3}{2}x^{-2/3} + C.$

7. $-\cos x + \sin x + C.$

8. $e^z + \ln|z| + C.$

9. $\frac{5}{7}x^{7/5} + C.$

10. $\int (x^2 + 4x)\,dx = \dfrac{x^3}{3} + 2x^2 + C = f(x);$

    $f(0) = 14 \Rightarrow C = 14.$ Hence

    $f(x) = \dfrac{x^3}{3} + 2x^2 + 14.$

> If one knows $f'(x)$, integration will produce a family of functions whose derivatives are $f(x)$. If one also knows a point on f, one can identify a specific member of the family.

11. $\int \cos x\,dx = \sin x + C = f(x); f\left(\dfrac{\pi}{2} \right) = 5 \Rightarrow$

    $C = 4.$ Hence $f(x) = \sin x + 4.$

12. $\int \dfrac{5}{x}\,dx = 5\ln|x| + C; f(e^2) = 5\ln e^2 + C =$

    $10 + C = 7.$ Hence $C = -3$ and

    $f(x) = 5\ln|x| - 3.$

13. $\int \sec^2 x\,dx = \tan x + C; f\left(\dfrac{\pi}{4} \right) = 1 + C =$

    $-3 \Rightarrow C = -4.$ Hence $f(x) = \tan x - 4.$

Section 5.2: Distance and Velocity from Acceleration with Initial Conditions

In Section 4.4, we obtained velocity and acceleration functions by successive differentiations of a distance function for motion along a line. In many instances, it is possible to use integration to do the reverse: that is, to produce velocity and distance functions if the acceleration function is known.

In the following examples and problems, t will be the time variable and distance, velocity, and acceleration functions will be represented by s, v, and a respectively, where

$$v(t) = s'(t) \quad \Rightarrow \quad s(t) = \int v(t)\, dt$$

$$a(t) = v'(t) \quad \Rightarrow \quad v(t) = \int a(t)\, dt.$$

Q1. A particle moving along a line has acceleration given by $a(t) = 4t$. It is known that $v(2) = 26$ and $s(1) = 12$. Find the velocity and distance functions. _____

> Answer: $v(t) = \displaystyle\int 4t\, dt = 2t^2 + C_1$; $v(2) = 26 = 8 + C_1 \Rightarrow C_1 = 18$. Hence $v(t) = 2t^2 + 18$.
>
> $s(t) = \displaystyle\int (2t^2 + 18)\, dt = \dfrac{2t^3}{3} + 18t + C_2$; $s(1) = 12 = \dfrac{2}{3} + 18 + C_2 \Rightarrow C_2 = -\dfrac{20}{3}$.
>
> Hence $s(t) = \dfrac{2t^3}{3} + 18t - \dfrac{20}{3}$.

Authors' Note: In working from acceleration to distance, two constants of integration are involved. One must have sufficient information to evaluate these constants in order to determine uniquely the distance function.

An object near the surface of the earth is attracted toward the earth by gravity sufficient to give it a downward acceleration of approximately 32 ft./sec.^{2} (or 9.8 m/sec^2).

When working with problems involving gravitational force, it is necessary to choose a coordinate system for reference. There are many possibilities, but it is often convenient to take the origin at the earth's surface and the positive direction to be upward. With this selection, acceleration due to gravity is -32 ft./sec.^{2} since it is in a negative direction relative to this coordinate system.

Q2. A ball is dropped from the top of a building 256′ high. How long will it take to reach the ground?

_____

> Answer: $a(t) = -32 \Rightarrow v(t) = \int (-32)\,dt = -32t + v_0$, where v_0 is the initial velocity. Since $v(0) = 0$, one obtains $v_0 = 0$; hence $v(t) = -32t$. Then $s(t) = \int (-32t)\,dt = -16t^2 + s_0$. Since $s(0) = 256$, one obtains $s_0 = 256$ and $s(t) = -16t^2 + 256$. The ball will reach the ground when $s(0) = 0$. Solving $-16t^2 + 256 = 0$, one obtains $t^2 = 16$, or $t = 4$ (seconds).

 Common error: Assuming that the integration constants v_0 and s_0 are always 0.

Problem Set 5.2

1. An object is hurled straight up with a velocity of 48 ft./sec.
   (a) When will the object reach its maximum height?
   (b) What is the maximum height?

2. From the top of a building 400 ft. high, a stone is dropped. Find when it will reach the ground and its impact velocity.

3. A particle moves along the x-axis in such a way that its acceleration at time t is $a(t) = 6t - 8$. The particle is at the origin when $t = 1$ and again when $t = 4$. Find
   (a) its velocity when it passes the origin
   (b) its velocity at $t = 2$
   (c) its distance from the origin at the start of its motion (at $t = 0$)

4. A boy is standing at the top of a cliff 160 ft. high.
   (a) If he throws a stone downward with a speed of 48 ft./sec., find how long it will take to hit the ground and its impact velocity.
   (b) Give the answers requested in (a) if the ball is thrown upward with a speed of 48 ft./sec.

5.

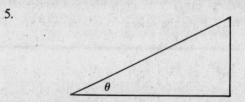

If a body moves on an incline making an angle of θ with the horizontal, then the gravitational acceleration is $32 \sin \theta$ ft./sec.^{2}. A ramp makes a $30°$ angle with the ground and a ball is rolled up the ramp with a speed of 56 ft./sec.
   (a) How far up the ramp will the ball go?
   (b) How high above the ground does the ball go?
   (c) How long does it take for the ball to go up and back?

Solutions and Comments for Problem Set 5.2

1. (a) $v(t) = -32t + 48$; $v(t) = 0$ when $t = 1\frac{1}{2}$ (sec.).
   (b) $s(t) = -16t^2 + 48t$; $s(1\frac{1}{2}) = -36 + 72 = 36$ (ft.).

2. $v(t) = -32t$, $s(t) = -16t^2 + 400$. $s(t) = 0$ when $t = 5$ (sec.). $v(5) = -160$ (ft./sec.).

> $v_0 = 48$ and $s_0 = 0$. In these problems, the positive direction will be upward and $s(t)$ will represent distance above the earth. Ball will reach max. height when its velocity is zero.

> Speed is considered to be the absolute value of velocity. The stone has a speed of 160 ft./sec. upon impact. The sign on the velocity value indicates direction (in this situation, downward).

3. $v(t) = 3t^2 - 8t + v_0$ and $s(t) = t^3 - 4t^2 + v_0 t + s_0$. $s(1) = 0 \Rightarrow v_0 + s_0 = 3$. $s(4) = 0 \Rightarrow 4v_0 + s_0 = 0$. Hence $-3v_0 = 3 \Rightarrow v_0 = -1$ and $s_0 = 4$. Hence $v(t) = 3t^2 - 8t - 1$ and $s(t) = t^3 - 4t^2 - t + 4$.

   (a) $v(1) = -6, v(4) = 15$;
   (b) $v(2) = -5$;
   (c) $s(0) = 4$.

4. (a) $v(t) = -32t - 48$ and $s(t) = -16t^2 - 48t + 160 = -16(t + 5)(t - 2)$. $s(t) = 0$ when $t = 2$; $v(2) = -112$ (ft./sec.).
   (b) $v(t) = -32t + 48$ and $s(t) = -16t^2 + 48t + 160 = -16(t - 5)(t + 2)$. $s(t) = 0$ when $t = 5$; $v(5) = -112$ (ft./sec.).

In (a), note that $t = -5$ is a solution for $s(t) = 0$, but -5 is not in the domain of t in this problem.

5. $a(t) = -32 \sin 30° = -32(\frac{1}{2}) = -16$.
   $v(t) = -16t + 56$; $s(t) = -8t^2 + 56t$.
   (a) $v(t) = 0$ when $t = 3\frac{1}{2}$; $s(3\frac{1}{2}) = 98$ (ft.).
   (b) $98 \sin 30° = 49$ (ft.).
   (c) $s(t) = 0$ when $-8t(t - 7) = 0$; $t = 7$ (sec.).

The ball will reach max. height when it stops rolling up the ramp; that is, when its velocity is zero.

Section 5.3: Solutions of $y' = ky$ and Applications to Growth and Decay

In many very realistic situations, the rate of change of a variable y is proportional to the variable itself. This can be expressed in the following way:

$$\frac{dy}{dt} = ky, \quad \text{where } k \text{ is the constant of proportionality.}$$

This is a simple differential equation and is sometimes referred to as the *law of natural growth and decay*. We can solve for y by "separating the variables," obtaining

$$\frac{dy}{y} = k\,dt,$$

and then integrating both sides of this equation with respect to each variable. The result is

$$\ln|y| = kt + k_1, \quad \text{where } k_1 \text{ is the constant of integration.}$$

Hence

$$y = e^{kt+k_1} = e^{kt}e^{k_1} = Ce^{kt}, \quad \text{where } C = e^{k_1}.$$

Authors' Note: We can drop the absolute value notation, since y is necessarily positive. Also, e^{k_1} is just a constant, so why not call it C?

Q1. A culture of bacteria triples itself every hour. What will be its percentage increase in 15 minutes?

Answer: If x_0 is the original number present and x is the number present at time t, then $x_0 = Ce^{k \times 0} = C$. Hence $x = x_0 e^{kt}$. Since $x = 3x_0$ when $t = 1$, one obtains $3x_0 = x_0 e^{k \cdot 1} \Rightarrow e^k = 3 \Rightarrow k = \ln 3$. Hence, $x = x_0 e^{t \cdot \ln 3} = x_0 e^{\ln 3^t} = x_0 \cdot 3^t$. When $t = \frac{1}{4}$, $x = x_0 \cdot 3^{\frac{1}{4}} = (1.316)x_0$. The increase is about 31.6% after 15 minutes.

Authors' Notes: We do not have to know the actual initial number to obtain the percentage increase.

Some individuals may prefer to start a problem of this type by writing $\frac{dx}{dt} = kx$, which is a symbolic statement of the actual situation. Note that this equation implies $\frac{dx}{x} = k\,dt \Rightarrow \ln x = kt + k_1$. Since $x = x_0$ when $t = 0$, one obtains $k_1 = \ln x_0$. Now $\ln x = kt + \ln x_0 \Rightarrow \ln x - \ln x_0 = kt \Rightarrow \ln\left(\frac{x}{x_0}\right) = kt \Rightarrow \frac{x}{x_0} = e^{kt} \Rightarrow x = x_0 e^{kt}$, which is the same result obtained above.

Common error: Confusing $\frac{dx}{dt} = kx$ and $\frac{dx}{dt} = kt$.

Authors' Note: If $\frac{dx}{dt} = kt$, then $x = k \times \frac{t^2}{2} + k_1$.

Problem Set 5.3

1. Express y in terms of x by separating the variables and solving the differential equation.

   (a) $\frac{dy}{dx} = 5y$, given that $y = 7$ when $x = 0$

   (b) $\frac{dy}{dx} = -10y$, given that $y = e^6$ when $x = 1$

2. A culture of yeast doubles itself every hour. What will be the percentage increase after three hours?

3. The rate of decay of a radioactive substance is proportional to the amount present. If half a gram of radium decays in 1600 years, what part of a gram will be left after 1000 years?

4. If x grams of a substance decay according to the formula $\frac{dx}{dt} = -0.04x$, how long will it take until only 50% of the original amount remains if the unit of time is 100 years?

5. Sugar in solution decomposes at a rate proportional to the amount present. If 35 pounds decomposes to 12 pounds in 3 hours, find when $\frac{1}{2}$ pound will remain.

Solutions and Comments for Problem Set 5.3

1. (a) $y = Ce^{5x}$; $7 = Ce^{5 \times 0} = C \times 1 \Rightarrow C = 7$. Hence $y = 7e^{5x}$.

   (b) $y = Ce^{-10x}$; $e^6 = Ce^{-10} \Rightarrow C = e^{16}$; Hence $y = e^{16}e^{-10x} = e^{16-10x}$.

2. $x = Ce^{kt}$; when $t = 0$, $x = x_0 \Rightarrow C = x_0$. Hence $x = x_0 e^{kt}$; when $t = 1$, $x = 2x_0 \Rightarrow e^k = 2 \Rightarrow k = \ln 2 \Rightarrow x = x_0 e^{t \ln 2} = x_0 e^{\ln 2^t} = x_0 \times 2^t$. When $t = 3$, we obtain $x = 8x_0$, a percentage increase of 700%.

> The alert reader should note that one does not need advanced mathematics for this problem. If something doubles every hour, there will be two of them after one hour, four of them after two hours, eight after three hours, etc.

$$t = \frac{\ln 2}{0.04} = \frac{0.69315}{0.04} = 17.33 \text{ (units of time)}.$$

(17.33)(100 years) = 1,733 years.

> Essentially, one must solve $\frac{1}{2}x_0 = x_0 e^{-0.04t}$ for t. Note that it is not necessary to know the actual amount of the original quantity.

3. $x = e^{\frac{-\ln 2}{1600}t} = e^{\ln 2^{-t/1600}} = 2^{-t/1600}$. When $t = 1000$, one obtains $x = 2^{-5/8} = 0.648$, or about 64.8%.

5. $x = 35e^{kt}$. When $t = 3$ and $x = 12$, we have $e^{3k} = \frac{12}{35} \Rightarrow k = \frac{1}{3}\ln\left(\frac{12}{35}\right) = \ln\left(\frac{12}{35}\right)^{t/3}$. Hence $x = 35e^{\ln(12/35)^{t/3}} = 35\left(\frac{12}{35}\right)^{t/3}$. When $x = \frac{1}{2}$, we obtain $\frac{1}{2} = 35\left(\frac{12}{35}\right)^{t/3} \Rightarrow \frac{1}{70} = \left(\frac{12}{35}\right)^{t/3} \Rightarrow t = 3\left[\frac{\ln 70}{\ln 35 - \ln 12}\right] = 11.9$ (hours).

> $x = x_0 e^{kt}$; $x = \frac{1}{2}x_0$ when $t = 1600 \Rightarrow$ $\frac{1}{2} = e^{1600k} \Rightarrow 1600k = \ln\left(\frac{1}{2}\right) = -\ln 2 \Rightarrow$ $k = \frac{-\ln 2}{1600}$. Keep in mind that an expression of the form $e^{\ln w}$ simplifies to w. This realization simplifies algebraic expressions relating to growth and decay.

> Knowledge of the laws of logarithms is essential for working these types of problems.

4. $x = x_0 e^{-0.04t} \cdot \frac{1}{2} = e^{-0.04t} \Rightarrow -0.04t = -\ln 2 \Rightarrow$

Section 5.4: Integration by Substitution

The basic formulas for integration are listed in Section 5.1. When the integration process for an integral is not immediately obvious, it may be possible to "reduce" the integral to one of the well-known forms by using a method of substitution.

Example 1: $\int \frac{(x+2)\,dx}{x^2 + 4x + 7}$: Letting $u = x^2 + 4x + 7$, then $du = (2x+4)\,dx = 2(x+2)\,dx$. With this substitution one obtains $\frac{1}{2}\int \frac{du}{u} = \frac{1}{2}\ln|u| + C = \frac{1}{2}\ln|x^2 + 4x + 7| + C$.

Example 2: $\int e^{\sin x}\cos x\,dx$: Letting $u = \sin x$, then $du = \cos x\,dx$ and one obtains $\int e^u\,du = e^u + C = e^{\sin x} + C$.

Authors' Note: Remember that one can check the result of an integration by differentiating the result. In Example 2 above, $\frac{d}{dx}(e^{\sin x} + C) = e^{\sin x}(\cos x)$, as expected.

Example 3: $\int \sqrt{5-x}\,dx$: Letting $u = 5 - x$, then $du = -dx$ and one obtains $-\int u^{1/2}\,du = -\frac{2}{3}u^{3/2} + C = $

$\frac{2}{3}(5-x)^{3/2} + C.$

Example 4: $\int \cos 7x\,dx$: Letting $u = 7x$, then $du = 7\,dx$ and one obtains $\frac{1}{7}\int \cos u\,du = \frac{1}{7}\sin u + C = $

$\frac{1}{7}\sin 7x + C.$

 Common error: $\int \cos 7x\,dx = \sin 7x + C.$

Example 5: $\int \dfrac{dx}{4x^2 + 9}$: Letting $u = 2x$, then $du = 2\,dx$ and one obtains $\frac{1}{2}\int \dfrac{du}{u^2 + 3^2} = \frac{1}{2} \times \left[\frac{1}{3}\tan^{-1}\frac{u}{3} + C_1\right]$

$= \frac{1}{6}\tan^{-1}\frac{2x}{3} + C.$

Authors' Note: There is no hard-and-fast rule about what substitution should be made. In general, one attempts to recognize "hidden forms" that will "surface" when a specific substitution is made. As is true with many things, proficiency comes with practice.

Q1. Integrate: (a) $\int \dfrac{4x^3\,dx}{x^4 + 5}$ _____   (b) $\int e^{x^2} x\,dx$ _____

Answers: (a) Letting $u = x^4 + 5 \Rightarrow du = 4x^3\,dx$ and the integral becomes $\int \dfrac{du}{u} = \ln|u| + C = $

$\ln(x^4 + 5) + C.$ Note $x^4 + 5 > 0.$

(b) Letting $u = x^2 \Rightarrow du = 2x\,dx$ and the integral becomes $\frac{1}{2}\int e^u\,du = \frac{1}{2}e^u + C = $

$\frac{1}{2}e^{x^2} + C.$

Problem Set 5.4

Perform the indicated integrations.

1. $\int \cos\frac{x}{3}\,dx$

2. $\int \dfrac{x^2\,dx}{x^3 + 2}$

3. $\int \dfrac{(4x + 6)\,dx}{x^2 + 3x + 2}$

4. $\int e^{\tan x} \sec^2 x\,dx$

5. $\int \sin^7 x \cos x \, dx$

6. $\int \dfrac{dx}{x^2 + 5}$

7. $\int \dfrac{dw}{\sqrt{25 - w^2}}$

8. $\int \tan^3 x \sec^2 x \, dx$

9. $\int \sec 5x \, dx$

10. $\int \dfrac{du}{u^2 - 36}$

Solutions and Comments for Problem Set 5.4

1. Letting $u = \dfrac{x}{3}$, we have $du = \dfrac{1}{3} dx$. Integral form is $\int 3 \cos u \, du$ and the desired integral is $3 \sin \dfrac{x}{3} + C$.

2. Letting $u = x^3 + 2$, we have $du = 3x^2 \, dx$. Desired integral is $\frac{1}{3} \ln |x^3 + 2| + C$.

3. Letting $u = x^2 + 3x + 2$, we have $du = (2x + 3) \, dx$. Desired integral is $2 \ln |x^2 + 3x + 2| + C$.

4. Letting $u = \tan x$, we have $du = \sec^2 x \, dx$ and the integral form is $\int e^u \, du$. The desired integral is $e^{\tan x} + C$.

5. Letting $u = \sin x$, we have $du = \cos x \, dx$. Integral form is $\int u^7 \, du$. Desired integral is $\dfrac{1}{8} \sin^8 x + C$.

6. $\dfrac{1}{\sqrt{5}} \tan^{-1} \dfrac{x}{\sqrt{5}} + C$.

> Write as $\dfrac{dx}{x^2 + (\sqrt{5})^2}$ and use formula (15) in Section 5.1.

7. $\sin^{-1} \dfrac{w}{5} + C$.

> Use (18) in Section 5.1 with $w = u$.

8. Letting $u = \tan x$, we have $du = \sec^2 x \, dx$ and the integral form is $\int u^3 \, du$. Desired integral is $\dfrac{1}{4} \tan^4 x + C$.

9. Using (7) in Section 5.1 with $u = 5x$, the desired integral is $\frac{1}{5} \ln |\sec 5x + \tan 5x| + C$.

10. Using (16) in Section 5.1 with $a = 6$, the desired integral is $\dfrac{1}{12} \ln \left| \dfrac{u - 6}{u + 6} \right| + C$.

*Section 5.4A: Integration by Trigonometric Substitution

We can often make a useful substitution for integration purposes by making use of the right triangle. When the integrand contains *(next page)*

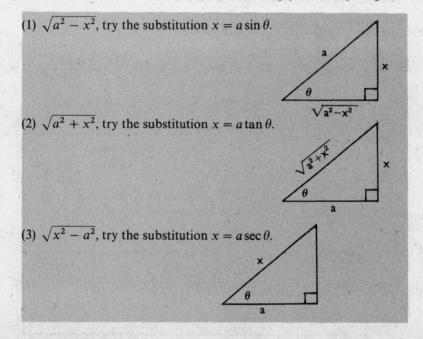

(1) $\sqrt{a^2 - x^2}$, try the substitution $x = a \sin \theta$.

(2) $\sqrt{a^2 + x^2}$, try the substitution $x = a \tan \theta$.

(3) $\sqrt{x^2 - a^2}$, try the substitution $x = a \sec \theta$.

Among other things, we can use these substitutions to derive integration formula (18) in Section 5.1.

Example: $\int \dfrac{dx}{\sqrt{a^2 - x^2}}$. Letting $x = a \sin \theta$, as suggested above, one obtains $dx = a \cos \theta \, d\theta$. Then the integral becomes $\int \dfrac{a \cos \theta \, d\theta}{\sqrt{a^2 - a^2 \sin^2 \theta}} = \int \dfrac{a \cos \theta \, d\theta}{\sqrt{a^2(1 - \sin^2 \theta)}} = \int \dfrac{a \cos \theta \, d\theta}{a \cos \theta} = \int 1 \, d\theta = \theta + C$. Since $x = a \sin \theta$, this implies $\sin \theta = \dfrac{x}{a}$ and $\theta = \sin^{-1} \dfrac{x}{a}$. Hence $\int \dfrac{dx}{\sqrt{a^2 - x^2}} = \sin^{-1} \dfrac{x}{a} + C$.

Problem Set 5.4A

Find

1. $\int \dfrac{dx}{(a^2 - x^2)^{3/2}}$

2. $\int \dfrac{dx}{(a^2 + x^2)^{3/2}}$

3. $\int \dfrac{dx}{(x^2 - a^2)^{3/2}}$

4. $\int \dfrac{du}{\sqrt{u^2 + a^2}}$

5. Evaluate $\int \dfrac{d\theta}{5 + 3 \cos \theta}$ by using the substitution $u = \tan \dfrac{\theta}{2}$.

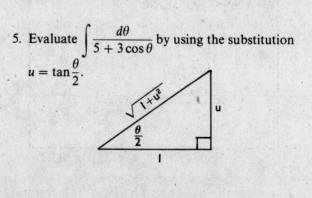

Solutions and Comments for Problem Set 5.4A

1. $x = a \sin \theta; \; dx = a \cos \theta \, d\theta.$ The integral becomes $\dfrac{1}{a^2} \displaystyle\int \sec^2 \theta \, d\theta = \dfrac{1}{a^2} \tan \theta + C$

$$= \dfrac{1}{a^2} \times \dfrac{x}{\sqrt{a^2 - x^2}} + C$$

$$= \dfrac{x}{a^2 \sqrt{a^2 - x^2}} + C.$$

With the indicated substitution,
$$(a^2 - x^2)^{3/2} = (a^2 - a^2 \sin^2 \theta)^{3/2}$$
$$= [a^2(1 - \sin^2 \theta)]^{3/2}$$
$$= (a^2 \cos^2 \theta)^{3/2}$$
$$= a^3 \cos^3 \theta.$$

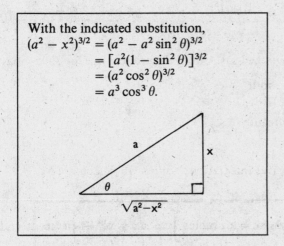

---

2. $x = a \tan \theta; \; dx = a \sec^2 \theta \, d\theta.$ The integral becomes $\dfrac{1}{a^2} \displaystyle\int \cos \theta \, d\theta = \dfrac{1}{a^2} \sin \theta + C$

$$= \dfrac{x}{a^2 \sqrt{x^2 + a^2}} + C.$$

With the indicated substitution,
$$(a^2 + x^2)^{3/2} = a^3 \sec^3 \theta.$$

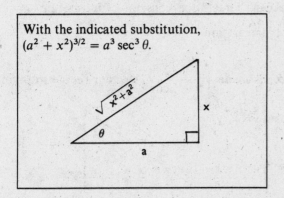

---

3. $x = a \sec \theta; \; dx = a \sec \theta \tan \theta \, d\theta.$ The integral becomes $\dfrac{1}{a^2} \displaystyle\int \dfrac{\cos \theta \, d\theta}{\sin^2 \theta}.$ If $u = \sin \theta$, then $du =$

$\cos \theta \, d\theta$ and one obtains $\dfrac{1}{a^2} \times \dfrac{(\sin \theta)^{-1}}{-1} + C =$

$$-\dfrac{1}{a^2} \times \dfrac{1}{\sin \theta} + C = -\dfrac{x}{a^2 \sqrt{x^2 - a^2}} + C.$$

With the indicated substitution,
$(x^2 - a^2)^{3/2} = a^3 \tan^3 \theta$ and the integral becomes $\dfrac{1}{a^2} \displaystyle\int \dfrac{\sec \theta \, d\theta}{\tan^2 \theta} = \dfrac{1}{a^2} \displaystyle\int \dfrac{\cos \theta \, d\theta}{\sin^2 \theta}.$

---

4. $u = a \tan \theta; \; du = a \sec^2 \theta \, d\theta.$ The integral becomes $\displaystyle\int \sec \theta \, d\theta = \ln |\sec \theta + \tan \theta| + C =$

$\ln \left| \dfrac{\sqrt{u^2 + a^2}}{a} + \dfrac{u}{a} \right| + C = \ln \left| u + \sqrt{u^2 + a^2} \right| -$

$\ln |a| + C = \ln \left| u + \sqrt{u^2 + a^2} \right| + C_1,$ where $C_1 = -\ln |a| + C.$

With the indicated substitution,
$$\sqrt{u^2 + a^2} = a \sec \theta.$$

---

5. $\cos \dfrac{\theta}{2} = \dfrac{1}{\sqrt{1 + u^2}}, \; \sin \dfrac{\theta}{2} = \dfrac{u}{\sqrt{1 + u^2}}.$ Hence

$\cos \theta = \cos^2 \dfrac{\theta}{2} - \sin^2 \dfrac{\theta}{2} = \dfrac{1 - u^2}{1 + u^2}.$ Also,

$du = \dfrac{1}{2} \sec^2 \dfrac{\theta}{2} d\theta \Rightarrow d\theta = 2 \cos^2 \dfrac{\theta}{2} du = \dfrac{2du}{1 + u^2}.$

The integral becomes $\displaystyle\int \dfrac{du}{u^2 + 4} = \dfrac{1}{2} \tan^{-1} \dfrac{u}{2} +$

$C = \dfrac{1}{2} \tan^{-1} \left(\dfrac{1}{2} \tan \dfrac{\theta}{2} \right) + C.$

There are many substitutions that can be made to evaluate integrals. As this example illustrates, the best substitution is sometimes far from obvious.

Section 5.5: Integration by Parts

An effective and versatile method of integration that can be applied to certain integrals is known as *integration by parts.*

$$\int f(x)g'(x)\,dx = f(x)g(x) - \int f'(x)g(x)\,dx$$

or, if $u = f(x)$ and $v = g(x)$,

$$\int u\,dv = uv - \int v\,du$$

Authors' Note: The formula for integration by parts is obtained easily from the product rule for differentiation. $\dfrac{d}{dx}(uv) = \dfrac{du}{dx}v + u\dfrac{dv}{dx}$. Integrating both sides yields $uv = \int v\dfrac{du}{dx}\,dx + \int u\dfrac{dv}{dx}\,dx = \int v\,du + \int u\,dv$, and the formula follows readily.

Q1. Use integration by parts to evaluate $\int xe^x\,dx$. _____

Answer: Let $u = x$ and $dv = e^x\,dx$. Then $du = dx$ and $v = e^x$. Hence $\int xe^x\,dx = xe^x - \int e^x\,dx$
$= xe^x - e^x + C$.

Authors' Note: If $dv = e^x\,dx$ then, to be exact, $v = e^x + C$. However, note that when applying integration by parts, $\int u\,dv = u(v + C) - \int (v + C)\,du = uv + uC - \int v\,du - \int C\,du = uv + uC$ $- \int v\,du - Cu = uv - \int v\,du$. Hence we can "forget" the constant of integration for v. However, don't forget the final constant of integration.

Problem Set 5.5

Evaluate

1. $\displaystyle\int x\ln x\,dx$

2. $\displaystyle\int \ln x\,dx$

3. $\displaystyle\int xe^{2x}\,dx$

4. $\displaystyle\int x\cos dx$

5. $\displaystyle\int w^n \ln w\,dw \quad$ if $n \neq -1$

6. $\displaystyle\int x\ln(x + 2)\,dx$

7. $\displaystyle\int \tan^{-1}x\,dx$

Solutions and Comments for Problem Set 5.5

1. $u = \ln x$ and $dv = xdx \Rightarrow du = \dfrac{dx}{x}$ and $v = \dfrac{x^2}{2}$.

   Desired integral is $\dfrac{x^2}{2}\ln x - \dfrac{x^2}{4} + C$.

> Note that if $n = -1$, then $\displaystyle\int w^{-1}\ln w\,dw = $
>
> $\displaystyle\int \ln w\,d(\ln w) = \dfrac{1}{2}(\ln w)^2 + C$.

2. $u = \ln x$ and $dv = dx \Rightarrow du = \dfrac{dx}{x}$ and $v = x$.

   Desired integral is $x \times \ln x - x + C$.

6. $u = \ln(x + 2)$ and $dv = xdx$. $du = \dfrac{dx}{x + 2}$ and

   $v = \dfrac{x^2}{2}$. The desired integral is

   $\dfrac{x^2}{2}\ln(x + 2) - \dfrac{x^2}{4} + x - 2\ln(x + 2) + C$.

3. $u = x$ and $dv = e^{2x}\,dx \Rightarrow du = dx$ and $v = \frac{1}{2}e^{2x}$.
   Desired integral is $\frac{1}{2}xe^{2x} - \frac{1}{4}e^{2x} + C$.

4. $u = x$ and $dv = \cos x\,dx \Rightarrow du = dx$ and $v = \sin x$. Desired integral is $x\sin x + \cos x + C$.

> In evaluating $\displaystyle\int \dfrac{x^2\,dx}{x + 2}$, it is helpful to write
>
> $\dfrac{x^2}{x + 2}$ as $x - 2 + \dfrac{4}{x + 2}$.

5. $u = \ln w$ and $dv = w^n\,dw \Rightarrow du = \dfrac{dw}{w}$ and $v = $

   $\dfrac{w^{n+1}}{n + 1}$. Hence one obtains $\dfrac{w^{n+1}}{n + 1}\ln w - $

   $\displaystyle\int \dfrac{w^{n+1}}{n + 1}\dfrac{dw}{w} = \dfrac{w^{n+1}}{n + 1}\ln w - \dfrac{w^{n+1}}{(n + 1)^2} + C$.

7. $u = \tan^{-1} x$ and $dv = dx \Rightarrow du = \dfrac{dx}{1 + x^2}$ and

   $v = x$. Desired integral is
   $x\tan^{-1} x - \frac{1}{2}\ln(x^2 + 1) + C$.

*Section 5.5A: Repeated Integration by Parts and Integral Reduction Formulas

It is sometimes necessary to apply the process of integration by parts more than once in order to evaluate a given integral.

Example:

$$\int e^x \cos x\,dx.$$

Let $u = e^x$ and $dv = \cos x\,dx$. Then $du = e^x\,dx$ and $v = \sin x$. Hence

$$(1) \quad \int e^x \cos x\,dx = e^x \sin x - \int e^x \sin x\,dx.$$

Now applying the process to $e^x \sin x\,dx$, let $u_1 = e^x$ and $dv_1 = \sin x\,dx$. Then $du_1 = e^x\,dx$ and $v_1 = -\cos x$ and one obtains

$$\int e^x \sin x\,dx = -e^x \cos x - \int (-\cos x)e^x\,dx = -e^x \cos x + \int e^x \cos x\,dx.$$

Substituting into (1) yields

$$\int e^x \cos x\, dx = e^x \sin x + e^x \cos x - \int e^x \cos x\, dx. \text{ Hence}$$

$$2 \int e^x \cos x\, dx = e^x \sin x + e^x \cos x \text{ and thus}$$

$$\int e^x \cos x\, dx = \frac{1}{2} e^x (\sin x + \cos x) + C. \quad \text{(Constant of integration has been added.)}$$

By writing $\int \cos^n x\, dx$ as $\int \cos^{n-1} x \cos x\, dx$ and letting $u = \cos^{n-1} x$, $dv = \cos x\, dx$, we can obtain the useful reduction formula

$$(2)\ n \int \cos^n x\, dx = \cos^{n-1} x \sin x + (n-1) \int \cos^{n-2} x\, dx.$$

In a similar manner, we can show

$$(3)\ n \int \sin^n x\, dx = -\sin^{n-1} x \cos x + (n-1) \int \sin^{n-2} x\, dx.$$

Problem Set 5.5A

In problems 1–5, evaluate the integral.

1. $\int x^2 e^x\, dx$

2. $\int x^3 e^x\, dx$

3. $\int e^u \sin u\, du$

4. $\int x^2 \ln x$

5. $\int \sec^3 \theta\, d\theta$

6. Use formula (2) to evaluate $\int \cos^2 x\, dx$.

7. Use formula (3) to evaluate $\int \sin^3 x\, dx$.

Solutions and Comments for Problem Set 5.5A

1. $u = x^2$ and $dv = e^x\, dx \Rightarrow du = 2x\, dx$ and $v = e^x$.
Then $\int x^2 e^x\, dx = x^2 e^x - \int 2xe^x\, dx =$
$e^x(x^2 - 2x + 2) + C.$

2. $u = x^3$ and $dv = e^x\, dx$. Then
$\int x^3 e^x\, dx = x^3 e^x - 3 \int x^2 e^x\, dx.$
Using the results from problem #1, we obtain
$e^x(x^3 - 3x^2 + 6x - 6) + C.$

3. $\frac{e^u}{2}(\sin u - \cos u) + C.$

> This is very similar to the example in the reading section.

4. $u = \ln x$ and $dv = x^2\, dx \Rightarrow du = \frac{1}{x}dx$ and $v = \frac{x^3}{3}.$

Then $\int x^2 \ln x \, dx = \dfrac{x^3}{3} \ln x - \int \dfrac{x^2}{3} dx =$

$\dfrac{x^3}{3} \ln x - \dfrac{x^3}{9} + C,$

$-\int \sec^3 \theta \, d\theta + \int \sec \theta \, d\theta.$ Solving for $\int \sec^3 \theta \, d\theta$

yields $\dfrac{1}{2}(\sec \theta \tan \theta + \ln |\sec \theta + \tan \theta|) + C.$

5. Write $\int \sec^3 \theta \, d\theta$ as $\int \sec \theta \, (\sec^2 \theta \, d\theta)$ and let

$u = \sec \theta$ and $dv = \sec^2 \theta \, d\theta$. Then $du =$
$\sec \theta \tan \theta \, d\theta$ and $v = \tan \theta$. Hence

$\int \sec^3 \theta \, d\theta = \sec \theta \tan \theta - \int \sec \theta \tan^2 \theta \, d\theta =$

$\sec \theta \tan \theta - \int \sec \theta \, (\sec^2 \theta - 1) \, d\theta = \sec \theta \tan \theta$

6. $2 \int \cos^2 x \, dx = \cos x \sin x + (2 - 1) \int 1 \, dx \Rightarrow$

$\int \cos^2 x \, dx = \dfrac{x}{2} + \dfrac{1}{2} \sin x \cos x + C.$

7. $3 \int \sin^3 x \, dx = -\sin^2 x \cos x + (3 - 1) \int \sin x \, dx$

$\Rightarrow \int \sin^3 x \, dx = -\dfrac{1}{3} \sin^2 x \cos x - \dfrac{2}{3} \cos x + C.$

*Section 5.6: Integration by Partial Fractions

When attempting to integrate a fractional function, it is often useful to write the function as a sum of fractions whose denominators are the factors of the original denominator. Any rational function $\dfrac{f(x)}{g(x)}$, where $f(x)$ and $g(x)$ are polynomials with the degree of $f(x)$ less than that of $g(x)$, can be written as a sum of its *partial fractions* related to the factors of $g(x)$ as follows:

(1) For each factor $x - a$ occurring only once in $g(x)$, there is a single fraction of the form $\dfrac{A}{x - a}$.

(2) For each factor $x - a$ occurring n times in $g(x)$ there is a group of fractions

$$\dfrac{A_1}{x - a} + \dfrac{A_2}{(x - a)^2} + \cdots + \dfrac{A_n}{(x - a)^n}.$$

(3) For each factor $x^2 + bx + c$ occurring only once in $g(x)$, there is a single fraction

$$\dfrac{Bx + C}{x^2 + bx + c}.$$

(4) For each factor $x^2 + bx + c$ occurring n times in $g(x)$ there is a group of fractions

$$\dfrac{B_1 x + C_1}{x^2 + bx + c} + \dfrac{B_2 x + C_2}{(x^2 + bx + c)^2} + \cdots + \dfrac{B_n x + C_n}{(x^2 + bx + c)^n}.$$

Example: $\dfrac{4x^4 - 1}{(x - 3)(x - 2)^2(x^2 + 4x + 7)^2} = \dfrac{A}{x - 3} + \dfrac{B}{x - 2} + \dfrac{C}{(x - 2)^2} + \dfrac{Dx + E}{x^2 + 4x + 7} + \dfrac{Fx + G}{(x^2 + 4x + 7)^2}$

Authors' Notes: The constants $A, B, C, D, E, F,$ and G can be found by clearing of fractions and equating like powers of x. (Algebraic manipulations are important in calculus.)

Note that linear and quadratic factors with leading coefficient $\neq 1$ can themselves be factored to obtain the desired form. For instance,

$$\dfrac{3x}{(2x + 1)(3x^2 + 6x + 2)} = \dfrac{3x}{[2(x + \frac{1}{2})][3(x^2 + 2x + \frac{2}{3})]} = \dfrac{\frac{1}{2}x}{(x + \frac{1}{2})(x^2 + 2x + \frac{2}{3})}.$$

Example: $\int \dfrac{2x^3 - x^2 - 14x + 11}{(x - 2)^2(x^2 + 1)}\,dx$ (Degree of numerator less than degree of denominator)

$$\text{Set}\quad \frac{2x^3 - x^2 - 14x + 11}{(x - 2)^2(x^2 + 1)} = \frac{A}{x - 2} + \frac{B}{(x - 2)^2} + \frac{Cx + D}{x^2 + 1}.$$

Clearing fractions, we obtain

$$2x^3 - x^2 - 14x + 11 = A(x^3 - 2x^3 + x - 2) + B(x^2 + 1) + C(x^3 - 4x^2 + 4x) + D(x^2 - 4x + 4).$$

Equating coefficients,

$$2 = A + C$$
$$-1 = -2A + B - 4C + D$$
$$-14 = A + 4C - 4D$$
$$11 = -2A + B + 4D.$$

Solving algebraically, $A = 2$, $B = -1$, $C = 0$, and $D = 4$. Hence

$$\int \frac{2x^3 - x^2 - 14x + 11}{(x - 2)^2(x^2 + 1)}\,dx = \int \left(\frac{2}{x - 2} - \frac{1}{(x - 2)^2} + \frac{4}{x^2 + 1} \right) dx$$

$$= 2\ln|x - 2| + \frac{1}{x - 2} + 4\tan^{-1} x + C.$$

The reader should take note that an integral like $\displaystyle\int \frac{dx}{x^2 + 2x + 7}$ often occurs when we are integrating using partial fractions. We can "handle" an integral like this by completing the square in the denominator and then using formula (15) in Section 5.1. Note that

$$\int \frac{dx}{x^2 + 2x + 7} = \int \frac{dx}{(x + 1)^2 + 6} = \frac{1}{\sqrt{6}} \tan^{-1}\left(\frac{x + 1}{\sqrt{6}} \right) + C.$$

Problem Set 5.6

Evaluate, using integration by partial fractions.

1. $\displaystyle\int \frac{2\,dx}{1 - x^2}$

2. $\displaystyle\int \frac{dw}{w^2 + w}$

3. $\displaystyle\int \frac{2x^2 + 5x - 1}{x(x - 1)(x + 2)}\,dx$

4. $\displaystyle\int \frac{4x^2 - 4}{x^3 + 2x^2}$

5. $\displaystyle\int \frac{12 - 8t}{t^2(t^2 + 4)}\,dt$

6. $\displaystyle\int \frac{3x^3\,dx}{(x - 1)^2(x^2 + x + 1)}$

Solutions and Comments for Problem Set 5.6

1. $-\ln|1 - x| + \ln|1 + x| + C = \ln\left| \dfrac{1 + x}{1 - x} \right| + C.$ 2. $\ln|w| - \ln|w + 1| + C = \ln\left| \dfrac{w}{w + 1} \right| + C.$

$$\frac{2}{(1 + x)(1 - x)} = \frac{A}{1 + x} + \frac{B}{1 - x} \Rightarrow$$
$$A = B = 1.$$

$$\frac{1}{w(w + 1)} = \frac{A}{w} + \frac{B}{w + 1} \Rightarrow A = 1,\ B = -1.$$

3. $\frac{1}{2}\ln|x| + 2\ln|x-1| - \frac{1}{2}\ln|x+2| + C.$

$$\boxed{\frac{2x^2 + 5x - 1}{x(x-1)(x+2)} = \frac{A}{x} + \frac{B}{x-1} + \frac{C}{x+2} \Rightarrow \\ A = \tfrac{1}{2}, B = 2, C = -\tfrac{1}{2}.}$$

$$\boxed{\frac{12 - 8t}{t^2(t^2 + 4)} = \frac{A}{t} + \frac{B}{t^2} + \frac{Ct+D}{t^2 + 4} \Rightarrow \\ A = -2, B = 3, C = 2, \text{ and } D = -3.}$$

4. $\ln|x| + \frac{2}{x} + 3\ln|x+2| + C.$

$$\boxed{\frac{4x^2 - 4}{x^2(x+2)} = \frac{A}{x} + \frac{B}{x^2} + \frac{C}{x+2} \Rightarrow \\ A = 1, B = -2, C = 3.}$$

6. $2\ln|x-1| - \frac{1}{x-1} + \frac{1}{2}\ln|x^2 + x + 1| + \frac{1}{\sqrt{3}}\tan^{-1}\left(\frac{2x+1}{\sqrt{3}}\right) + C.$

$$\boxed{\frac{3x^3}{(x-1)^2(x^2 + x + 1)} = \frac{A}{x-1} + \frac{B}{(x-1)^2} \\ + \frac{Cx+D}{x^2 + x + 1} \Rightarrow A = 2, B = 1, C = 1, \\ D = 1.}$$

5. $-2\ln|t| - \frac{3}{t} + \ln(t^2 + 4) - \frac{3}{2}\tan^{-1}\frac{t}{2} + C.$

Section 5.7: Multiple Choice Questions for Chapter 5

1. $\displaystyle\int (x^5 + x^{1/2})\,dx =$
    (A) $x^6 + \frac{1}{2}x^{-1/2} + C$
    (B) $\frac{x^6}{6} + x^{-1/2} + C$
    (C) $\frac{x^6}{6} + \frac{1}{2}x^{-1/2} + C$
    (D) $5x^4 + \frac{1}{2}x^{-1/2} + C$
    (E) None of the answer choices is correct.

2. If $w'(x) = 4x^3 + 3x^2 + 4x$ and $w(1) = 12$, then $w(x) =$
    (A) $12x^2 + 6x + 4$
    (B) $x^4 + x^3 + 2x^2 + 8$
    (C) $x^4 + 3x^3 + 4x^2 + 4$
    (D) $\frac{x^4}{4} + \frac{x^3}{3} + 4x^2 + 6$
    (E) None of the answer choices is correct.

3. If $f'(x) = \frac{3}{x}$ and $f(e^5) = 3$, then $f(x) =$
    (A) $3x^{-1} + 2$
    (B) $\ln|x| + 3$
    (C) $\ln|x| + e^7$
    (D) $3(\ln|x|) - 12$
    (E) None of the answer choices is correct.

4. A particle moving along a line has acceleration given by $a(t) = 3t^2$. At $t = 2$, the velocity of the particle is 20 and its distance from the origin is 34. What is the distance function?
    (A) $s(t) = t^4 + 12t + 18$
    (B) $s(t) = t^3 + 12$
    (C) $s(t) = t^3 + 12t + 6$
    (D) $s(t) = \frac{t^4}{4} + 12t + 6$
    (E) $s(t) = t^4 - 6t^2 + 3t + 2$

5. A stone is hurled downward from the top of a 192 ft. building with an initial speed of 64 ft./sec. What is its speed when it hits the ground?
    (A) 64 ft./sec.
    (B) 128 ft./sec.
    (C) 160 ft./sec.
    (D) 48 ft./sec.
    (E) 96 ft./sec.

6. $\displaystyle\int 3xe^{4x}\,dx =$
    (A) $xe^{4x} - e^{4x} + C$
    (B) $\frac{x}{5}e^{5x} - \frac{1}{4}e^{4x} + C$
    (C) $\frac{3}{4}xe^{4x} - \frac{3}{16}e^{4x} + C$
    (D) $\frac{3}{8}e^{4x} - xe^{4x} + C$
    (E) None of the answer choices is correct.

7. $\displaystyle\int \frac{x+2}{\sqrt{x}}\,dx =$

(A) $\frac{2}{3}x^{3/2} + 4x^{1/2} + C$

(B) $(x+2)^2 + C$

(C) $\frac{1}{2}x^{-1/2} + 2x^{1/2} + C$

(D) $2x^{3/2} + x^{-1/2} + C$

(E) None of the answer choices is correct.

8. $\displaystyle\int \frac{3\,dx}{x^2 + 16} =$

(A) $\tan^{-1}\left(\frac{x}{4}\right) + C$

(B) $\tan^{-1}\left(\frac{x}{16}\right) + C$

(C) $\frac{3}{4}\tan^{-1}\left(\frac{x}{4}\right) + C$

(D) $\frac{1}{4}\sec\left(\frac{x}{4}\right) + C$

(E) None of the answer choices is correct.

9. $\displaystyle\int \frac{x\,dx}{x+3} =$

(A) $\frac{1}{2}(x+3)^{-1/2} + C$

(B) $\ln|x+3| + C$

(C) $x - \ln|x+3| + c$

(D) $x - 3\ln|x+3| + C$

(E) None of the answer choices is correct.

10. $\displaystyle\int x\cos 3x\,dx =$

(A) $\frac{x}{3}\sin 3x + \frac{1}{9}\cos 3x + C$

(B) $\sin 3x + x\cos 3x + C$

(C) $\frac{x}{3}\cos 3x + \sin 3x + C$

(D) $x\sin 3x + \cos 3x + C$

(E) None of the answer choices is correct.

11. If $\dfrac{dy}{dx} = 6y$ and $y = e^4$ when $x = 3$, then $y =$

(A) $e^{4x/3}$

(B) $6e^{6x}$

(C) e^{4x}

(D) $6e^{7-x}$

(E) e^{6x-14}

12. $\displaystyle\int e^{\sin x}\cos x\,dx =$

(A) $e^{\cos x} + \sin x + c$

(B) $e^{\cos x} + C$

(C) $e^{\sin x} + C$

(D) $e^{\cos x}(\sin x) + C$

(E) None of the answer choices is correct.

13. A culture of bacteria doubles itself every hour. If x_0 is the original number present, then the amount present after 1 hour and 40 minutes is

(A) $\frac{5}{3}x_0$

(B) $\sqrt[3]{32}x_0$

(C) $\frac{9}{2}x_0$

(D) $\sqrt{2}x_0$

(E) $\sqrt{8}x_0$

14. $\displaystyle\int (\sec^2 x + \sec x\tan x)\,dx =$

(A) $\tan x + \csc x + C$

(B) $\tan x + \sec x + C$

(C) $\dfrac{\sec^3 x}{3} + \tan^2 x + C$

(D) $\ln|\sec x + \tan x| + C$

(E) None of the answer choices is correct.

15. A function f is such that $f'(x) = 2x + 3$. The equation of the tangent line at the point where $x = 1$ is $y - 11 = 5(x - 1)$. What is the function f?

(A) $f(x) = 2x^2 + 4x + 5$

(B) $f(x) = x^2 + 3x + 7$

(C) $f(x) = 5x^2 + 6$

(D) $f(x) = x^2 + 3x + 11$

(E) $f(x) = x^2 + 10x$

16. $\displaystyle\int \frac{e^{\tan^{-1} x}}{x^2 + 1} =$

(A) $e^{\tan^{-1} x} + C$

(B) $e^x[\ln(x^2 + 1)] + C$

(C) $(2x + 1)e^{\tan^{-1} x} + C$

(D) $2xe^{\tan^{-1} x} + C$

(E) None of the answer choices is correct.

17. $\displaystyle\int \frac{x^2}{2x^3 - 4} =$

(A) $\dfrac{x^4}{4} - 4x + C$

(B) $\ln|2x^3 - 4| + C$

(C) $\frac{1}{6} \times \ln|2x^3 - 4| + C$

(D) $e^{2x^3 - 4} + C$

(E) None of the answer choices is correct.

*18. $\displaystyle\int \frac{dx}{\sqrt{25 - x^2}} =$

(A) $2x(25 - x^2)^{-1/2} + C$

(B) $\frac{x}{2}(25 - x^2)^{-3/2} + C$

(C) $\tan^{-1}\frac{x}{5} + C$

(D) $\sin^{-1}\frac{x}{5} + C$

(E) None of the answer choices is correct.

*19. $\displaystyle\int \frac{dx}{(x-3)(x+4)} =$

(A) $\ln|(x-3)(x+4)| + C$

(B) $\frac{1}{7}\ln|(x-3)(x+4)| + C$

(C) $\frac{2}{7}\ln\left|\dfrac{x+4}{x-3}\right| + C$

(D) $\frac{1}{7}\ln\left|\dfrac{x-3}{x+4}\right| + C$

(E) None of the answer choices is correct.

*20. $\displaystyle\int e^{2x}\cos 3x\, dx =$

(A) $\dfrac{e^{2x}}{13}(3\sin 3x + 2\cos 3x) + C$

(B) $\dfrac{e^{2x}}{13}(3\cos 3x + 2\sin 3x) + C$

(C) $2e^{2x}\sin 3x + C$

(D) $e^{2x}\sin 3x + 2e^{2x}\cos 3x + C$

(E) None of the answer choices is correct.

Solutions and Comments for Multiple Choice Questions

1. **(E)** The integral is $\dfrac{x^6}{6} + \dfrac{2}{3}x^{3/2} + C$.

2. **(B)** $w(x) = x^4 + x^3 + 2x^2 + C$; $w(1) = 12 \Rightarrow$
   $C = 8$.

> Note that answer choices (A), (C), and (D) could be eliminated since they don't produce the given expression $w'(x)$.

3. **(D)** $f(x) = 3(\ln|x|) + C.\ f(e^5) = 3 \Rightarrow$
   $C = -12$.

4. **(D)** $v(t) = t^3 + C_1$; $v(2) = 20 \Rightarrow C_1 = 12$.
   $s(t) = \dfrac{t^4}{4} + 12t + C_2$; $s(2) = 34 \Rightarrow C_2 = 6$.

5. **(B)** $a(t) = -32$; $v(t) = -32t + v_0$;
   $v(0) = -64$ (downward) $\Rightarrow v_0 = -64$. If s is
   distance above ground, then $s(0) = 192$. Hence
   $s(t) = -16t^2 - 64t + 192 =$
   $-16(t+6)(t-2) = 0$
   when $t = 2$. $v(2) = -128 \Rightarrow$ impact speed $=$
   128 (ft./sec.).

6. **(C)** Use integration by parts, letting $u = 3x$
   and $dv = e^{4x}\,dx$. Then $du = 3\,dx$ and $v = \frac{1}{4}e^{4x}$
   and the integral becomes $\dfrac{3}{4}xe^{4x} - \dfrac{3}{4}\displaystyle\int e^{4x}\,dx.$

7. **(A)** Write as $\displaystyle\int (x^{1/2} + 2x^{-1/2})\,dx.$

8. **(C)** Write as $\dfrac{3}{16}\displaystyle\int \frac{1}{(x^2/16)+1}\,dx =$
   $\dfrac{3}{4}\displaystyle\int \frac{1}{(x/4)^2+1}\left(\frac{1}{4}dx\right)$ and use formula (15) in
   Section 5.1.

9. **(D)** Write as
   $$\int \frac{(x+3)-3}{x+3}\,dx = \int\left(1 - \frac{3}{x+3}\right)dx.$$

10. **(A)** Use integration by parts, letting $u = x$ and
    $dv = \cos 3x\,dx$. Then $du = dx$ and $v = \frac{1}{3}\sin 3x$
    and the integral becomes
    $$\frac{x}{3}\sin 3x - \int \frac{1}{3}\sin 3x\,dx.$$

11. **(E)** Write as $\dfrac{dy}{y} = 6\,dx$. Then $\ln|y| = 6x + C$
    and $y = e^{6x+C} = e^{6x}e^C$. Since $y = e^4$ when
    $x = 3$, we have $e^C = e^{-14}$.

> Some of the answer choices could be eliminated by checking to see if $x = 3$ produces a y value equal to e^4.

12. **(C)** If $u = \sin x$, then $du = \cos x\,dx$ and we
    have the form $\displaystyle\int e^u\,du.$

13. **(B)** $x = x_0 2^t$. When $t = \frac{5}{3}$ hrs., we have
    $x = x_0 2^{5/3} = \sqrt[3]{2^5}x_0.$

> $x = x_0 e^{kt}$. When $t = 1$, $x = 2x_0 \Rightarrow e^k = 2 \Rightarrow$
> $k = \ln 2$. Hence $x = x_0 e^{t(\ln 2)} = x_0 2^t.$

14. **(B)** A direct application of formulas (9) and
    (11) in Section 5.1.

15. **(B)** $f(x) = x^2 + 3x + C$. Since the tangent
    line intersects f at $(1, 11)$, we have $C = 7$.

16. **(A)** If $u = \tan^{-1} x$, then $du = \dfrac{1}{x^2 + 1} dx$ and the integral has the familiar form $\displaystyle\int e^u \, du$.

17. **(C)** If $u = 2x^3 - 4$, then $du = 6x^2 \, dx$ and the integral form is $\dfrac{1}{6} \displaystyle\int \dfrac{1}{u} du = \dfrac{1}{6} \times \ln|u| + C$.

18. **(D)** Use the substitution $x = 5 \sin \theta$. (See Section 5.4A.)

19. **(D)** Use partial fractions, writing
$$\frac{1}{(x - 3)(x + 4)} = \frac{A}{x - 3} + \frac{B}{x + 4} =$$

$$\frac{(A + B)x + (4A - 3B)}{(x - 3)(x + 4)},$$ we obtain $A = 1/7$ and $B = -1/7$. Integral becomes $\dfrac{1}{7} \displaystyle\int \dfrac{dx}{x - 3} - \dfrac{1}{7} \displaystyle\int \dfrac{dx}{x + 4} = \dfrac{1}{7}(\ln|x - 3| - \ln|x + 4|) + C.$

20. **(A)** This requires integrating by parts twice. If $u = e^{2x}$ and $dv = \cos 3x \, dx$, then the integral becomes $\dfrac{1}{3} e^{2x} \sin 3x - \dfrac{2}{3} \displaystyle\int e^{2x} \sin 3x \, dx.$ On the second application, let $u = e^{2x}$ and $dv = \sin 3x \, dx$. Then $du = 2e^{2x} \, dx$ and $v = -\frac{1}{3} \cos 3x.$

Chapter 6
THE DEFINITE INTEGRAL

Section 6.1: The Definite Integral as an Area

If f is a function continuous on a closed interval $[a,b]$, we can divide the interval into n equal subintervals of length $\dfrac{b-a}{n}$. Referring to the figure, the endpoints of these subintervals are

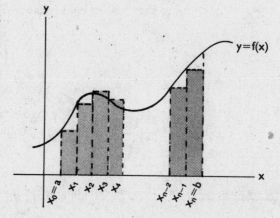

$$x_0 = a$$
$$x_1 = a + \Delta x$$
$$x_2 = a + 2\Delta x$$
$$\cdots$$
$$x_k = a + k\Delta x$$
$$\cdots$$
$$x_n = a + n\Delta x = b.$$

Considering the situation where $f(x) > 0$ on $[a,b]$, if we let the symbolism $\displaystyle\int_a^b f(x)\,dx$ represent the area bounded by $y = f(x)$, $x = a$, $x = b$, and the x-axis, then it should be intuitively obvious that we can "approximate" this area by adding up the areas of the n rectangles whose altitudes are equal to $f(x_0), f(x_1), f(x_2), \ldots, f(x_{n-1})$. Using the symbol $\approx$ to mean "is approximately equal to," we have

$$(1) \quad \int_a^b f(x)\,dx \approx f(x_0)\Delta x + f(x_1)\Delta x + \cdots + f(x_{n-1})\Delta x = \Delta x \sum_{k=0}^{n-1} f(x_k).$$

The reader should note that we could use a different set of rectangles to approximate $\displaystyle\int_a^b f(x)\,dx$.

$$(2) \quad \int_a^b f(x)\,dx \approx \Delta x \sum_{k=1}^{n} f(x_k).$$

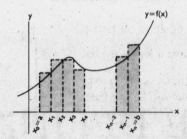

Formulas (1) and (2) represent the *rectangular rule* for approximating $\int_a^b f(x)\,dx$.

The larger the value of n, the closer the computations produced by the rectangular rule will be to the actual area $\int_a^b f(x)\,dx$.

Authors' Note: $\int_a^b f(x)\,dx$, which is called the *definite integral* of $f(x)$ between a and b, is just symbolism at the present time and will be elaborated upon in the following sections. However, it can be noted that this symbolism can be related to the process of adding up the areas of rectangles.

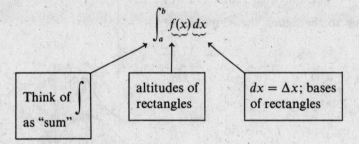

$$\int_a^b f(x)\,dx$$

| Think of $\int$ as "sum" | altitudes of rectangles | $dx = \Delta x$; bases of rectangles |

Q1. Approximate $\int_0^1 x^2\,dx$ by dividing $[0,1]$ into four subintervals and (a) using formula (1); (b) using formula (2).

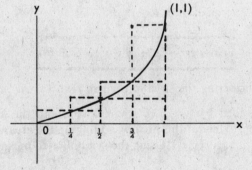

Answers: (a) $\Delta x \times \sum_{k=0}^{3} f(x_k) = \dfrac{1}{4}\left[0^2 + \left(\dfrac{1}{4}\right)^2 + \left(\dfrac{1}{2}\right)^2 + \left(\dfrac{3}{4}\right)^2\right] = \dfrac{1}{4} \times \dfrac{14}{16} = \dfrac{7}{32}.$

(b) $\Delta x \times \sum_{k=1}^{4} f(x_k) = \dfrac{1}{4}\left[\left(\dfrac{1}{4}\right)^2 + \left(\dfrac{1}{2}\right)^2 + \left(\dfrac{3}{4}\right)^2 + 1^2\right] = \dfrac{1}{4} \times \dfrac{30}{16} = \dfrac{15}{32}.$

Authors' Note: Noting the curvature of $y = x^2$ on $[0,1]$, we can realize that $\dfrac{7}{32} < \int_0^1 x^2\,dx < \dfrac{15}{32}$. Also, since 4 is not a large number of subintervals, the approximations should not be expected to be very accurate. The actual area is $\frac{1}{3}$ (sq. units).

If $f(x)$ is continuous, but not always positive on $[a, b]$, then $\int_a^b f(x)\,dx$ does not necessarily have a straight area interpretation, but it can be related to the concept of area. In the figure, if we consider the interval $[a, c]$, then $\Delta x \times f(x_k) \geq 0$ since $\Delta x > 0$ and $f(x_k) \geq 0$ for $x_k \in [a, c]$. However, in $[c, b]$, $\Delta x \times f(x_j) \leq 0$ since $\Delta x > 0$ and $f(x_j) \leq 0$ for $x_j \in [c, b]$. Hence, it should not be surprising to realize that if A and B represent the areas of the shaded regions in the figure below, then $\int_a^b f(x) = A - B$.

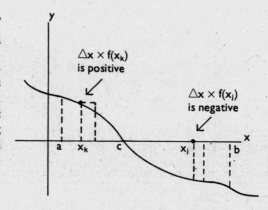

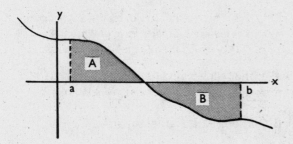

Q2. If A, B, C, D and E are the areas of the shaded regions, express $\int_a^b f(x)\,dx$ in terms of these letters.

Answer: $-A + B - C + D - E$.

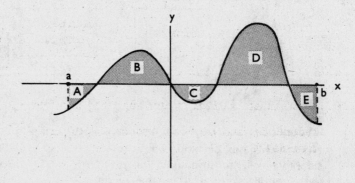

Problem Set 6.1

1. The capital letters represent actual areas. Express $\int_a^b f(x)\,dx$ in terms of these letters.

(a)

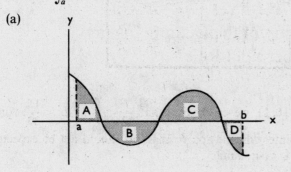

(b)

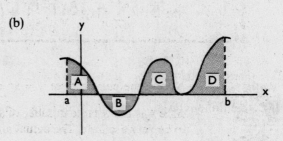

(c)

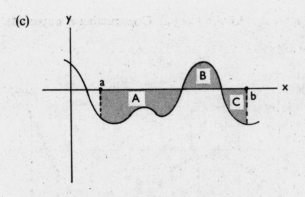

2. Approximate $\int_0^1 x^2 \, dx$ by letting $\Delta x = \dfrac{1}{10}$ and using (a) formula (1); (b) formula (2).

3. Use formulas (1) and (2) to approximate $\int_1^2 \dfrac{dx}{1 + x^2}$ by dividing $[1, 2]$ into 4 subintervals.

4. Use formulas (1) and (2) to approximate $\int_0^2 (x^2 - 1) \, dx$ by dividing $[0, 2]$ into 8 subintervals.

5. Show that $\Delta x \times \sum\limits_{k=1}^{n} f(x_k) - \Delta x \times \sum\limits_{k=0}^{n-1} f(x_k)$

$$\Delta x = [f(x_n) - f(x_0)].$$

Note that this result could be used to check the two calculations made in problems 2-4.

Solutions and Comments for Problem Set 6.1

1. (a) $A - B + C - D$;
   (b) $A - B + C + D$;
   (c) $-A + B - C$.

2. $\dfrac{1}{10}\left[0^2 + \left(\dfrac{1}{10}\right)^2 + \cdots + \left(\dfrac{9}{10}\right)^2\right]$

   $= \dfrac{1}{10} \times \dfrac{285}{100} = 0.285$;

   (b) $\dfrac{1}{10}\left[\left(\dfrac{1}{10}\right)^2 + \left(\dfrac{2}{10}\right)^2 + \cdots + 1^2\right]$

   $= \dfrac{1}{10} \times \dfrac{385}{100} = 0.385$.

> Note that these approximations are considerably closer to the actual area 1/3 than those obtained in Q1 in the reading section.

> If you look at the graph of $y = \dfrac{1}{1 + x^2}$, it can be seen why formula (1) produces a larger value than formula (2).

4. $\dfrac{1}{4}\Bigg[-1 + \left(-\dfrac{15}{16}\right) + \left(-\dfrac{3}{4}\right) + \left(-\dfrac{7}{16}\right) + 0 +$

   $\dfrac{9}{16} + \dfrac{20}{16} + \dfrac{33}{16}\Bigg] = \dfrac{1}{4} \times \dfrac{12}{16} = 0.1875$.

   Formula (2) produces 1.1875.

> The two approximations are not "close" to one another. Note that the number of subintervals is small; hence one should not expect too much accuracy.

3. $\dfrac{1}{4}\left[\dfrac{1}{2} + \dfrac{16}{41} + \dfrac{16}{52} + \dfrac{16}{65}\right] = 0.361$;

   $\dfrac{1}{4}\left[\dfrac{16}{41} + \dfrac{16}{52} + \dfrac{16}{65} + \dfrac{1}{5}\right] = 0.286$.

5. This result follows readily. Note that if we use the result to check the values obtained in problem #4, we have $\dfrac{1}{4}[f(2) - f(0)] = \dfrac{1}{4}[3 - (-1)] = \dfrac{1}{4} \times 4 = 1$.

*Section 6.1A: Approximation of the Definite Integral Using the Trapezoidal Rule

Another method for approximating the definite integral $\int_a^b f(x) \, dx$ is the *trapezodial rule*.

If f is continuous on $[a, b]$, we can divide the interval in n equal subintervals of length $\dfrac{b - a}{n} = \Delta x$. If $f(x) > 0$

on $[a, b]$, then we have a straight area interpretation for the definite integral. Constructing n trapezoids, as shown in the figure, we can note that

Area of first trapezoid $= \frac{1}{2}[f(x_0) + f(x_1)]\Delta x$,

Area of second trapezoid $= \frac{1}{2}[f(x_1) + f(x_2)]\Delta x$, ...

Area of n-th trapezoid $= \frac{1}{2}[f(x_{n-1}) + f(x_n)]\Delta x$.

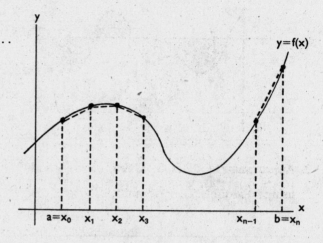

Hence

$$\int_a^b f(x)\, dx \approx \left[\frac{1}{2}f(x_0) + f(x_1) + f(x_2) + \cdots + f(x_{n-1}) + \frac{1}{2}f(x_n)\right]\Delta x = \left[\frac{1}{2}f(a) + \sum_{k=1}^{n-1} f(x_k) + \frac{1}{2}f(b)\right]\Delta x.$$

Q1. Find an approximation for $\int_1^5 x^2\, dx$ using the trapezodial rule and dividing $[1, 5]$ into four equal subintervals.

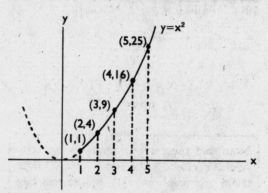

Answer: $\Delta x = 1$. The approximate area is
$$\left[\tfrac{1}{2} \times 1 + 4 + 9 + 16 + \tfrac{1}{2} \times 25\right] \times 1 = 42 \text{ (sq. units)}.$$

Authors' Note: The actual area is $41\frac{1}{3}$ sq. units. The curvature of the function on $[1, 5]$ should make it obvious why the approximation is greater than the actual area.

Problem Set 6.1A

Use the trapezoidal rule to approximate the value of the definite integral.

1. $\int_1^{10} x^2\, dx$ with $n = 9$

2. $\int_0^1 \frac{dx}{1 + x^2}$ with $n = 4$

3. $\int_0^{10} x^3\, dx$ with $n = 10$

4. $\int_{-1}^{3} \sqrt{4 + x^3}\, dx$ with $n = 4$ 5. $\int_{1}^{2} \frac{1}{x}\, dx$ with $n = 10$

Solutions and Comments for Problem Set 6.1A

1. $[\frac{1}{2} + 4 + 9 + 16 + 25 + 36 + 49 + 64 + 81 + \frac{1}{2}(100)] \times 1 = 334\frac{1}{2}$.

> The actual area is 2500 (sq. units).

> The actual area is 333 (sq. units).

4. $[\frac{1}{2}\sqrt{3} + 2 + \sqrt{5} + \sqrt{12} + \frac{1}{2}\sqrt{31}] \times 1 = 11.35$.

> There is no easy way to actually evaluate the indicated integral.

2. $[0.5 + 0.9412 + 0.8000 + 0.6400 + 0.250] \times (0.25) = 0.7828$.

> The actual area, to four significant figures, is 0.7854 (sq. units).

5. $\left[\frac{1}{2} + \frac{1}{1.1} + \frac{1}{1.2} + \frac{1}{1.3} + \frac{1}{1.4} + \frac{1}{1.5} + \frac{1}{1.6} + \frac{1}{1.7} + \frac{1}{1.8} + \frac{1}{1.9} + \frac{1}{4}\right] \times (0.1) = 0.6938$.

> The actual area, to four significant figures, is 0.6931 (sq. units).

3. $[0 + 1^3 + 2^3 + 3^3 + 4^3 + 5^3 + 6^3 + 7^3 + 8^3 + 9^3 + \frac{1}{2}10^3] \times 1 = 2{,}525$.

•   Section 6.2: The Definite Integral as the Limit of a Sum

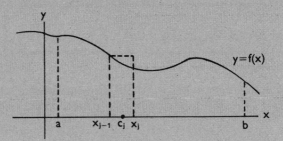

If (a) f is continuous on $[a, b]$;

  (b) $[a, b]$ is divided into n subintervals of length $\dfrac{b - a}{n} = \Delta x$;

  (c) c_j is any point in the j-th subinterval; that is, $c_j \in [x_{j-1}, x_j]$ where $j = 1, 2, 3, \ldots, n$

then $\displaystyle\int_{a}^{b} f(x)\, dx \approx \Delta x \times \sum_{j=1}^{n} f(c_j)$.

Authors' Note: In Section 6.1, c_j was either x_{j-1} or x_j; that is, c_j was an endpoint of a subinterval. Geometrically speaking, it should be intuitively obvious that c_j could be any point in $[x_{j-1}, x_j]$.

The exact value of $\int_a^b f(x)\,dx$ is $\lim\limits_{n\to\infty} \Delta x \times \sum\limits_{j=1}^{n} f(c_j) = \lim\limits_{n\to\infty} \dfrac{b-a}{n} \times \sum\limits_{j=1}^{n} f(c_j)$. We will now relate the symbolism $\int_a^b f(x)\,dx$ to the material on the indefinite integral in Section 6.1.

> **The Fundamental Theorem of Integral Calculus:**
> If f is continuous on $[a, b]$ and if $F' = f$, then
>
> $$\lim_{n\to\infty} \frac{b-a}{n} \times \sum_{j=1}^{n} f(c_j) = \int_a^b f(x)\,dx = F(b) - F(a).$$

When evaluating definite integrals, the notation $F(x)\Big|_a^b$ is used to represent $F(b) - F(a)$.

Q1. Evaluate $\int_2^5 x^2\,dx.$ _____

> Answer: $\displaystyle\int_2^5 x^2\,dx = \frac{x^3}{3}\bigg|_2^5 = \frac{5^3}{3} - \frac{2^3}{3} = \frac{117}{3} = 39.$

Authors' Note: When evaluating the definite integral, the constant of integration is not relevant since it "subtracts itself out." Note that

$$\int_b^a f(x)\,dx = (F(x) + C)\bigg|_a^b = [F(b) + C] - [F(a) + C] = F(b) - F(a).$$

We should not forget that $\int_a^b f(x)\,dx$ is the limit of a sum as the number of terms increases indefinitely. Sometimes it is possible to "unmask" a complicated-looking sum and evaluate it as a definite integral.

Q2. Evaluate $\lim\limits_{n\to\infty} \dfrac{1}{n} \times \sum\limits_{k=1}^{n} \left(\dfrac{k}{n}\right)^2.$ _____

> Answer: $\displaystyle\int_0^1 x^2\,dx = \frac{x^3}{3}\bigg|_0^1 = \frac{1}{3} - 0 = \frac{1}{3}.$

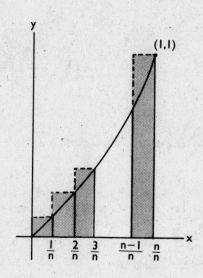

Authors' Note: In Q2, let $\Delta x = \dfrac{1}{n}$ and note that $\dfrac{1}{n} \times \sum\limits_{k=1}^{n} \left(\dfrac{k}{n}\right)^2 =$ $\dfrac{1}{n}\left(\dfrac{1}{n}\right)^2 + \dfrac{1}{n}\left(\dfrac{2}{n}\right)^2 + \cdots + \dfrac{1}{n}\left(\dfrac{n}{n}\right)^2$. Thinking of the geometry of the situation, we are adding up the areas of rectangles with base $\dfrac{1}{n}$ and altitudes $\left(\dfrac{1}{n}\right)^2$, $\left(\dfrac{2}{n}\right)^2, \ldots, \left(\dfrac{n}{n}\right)^2$.

Problem Set 6.2

In problems 1–9, evaluate the integral.

1. $\displaystyle\int_0^1 x^3\,dx$

2. $\displaystyle\int_2^4 \frac{1}{x^2}\,dx$

3. $\displaystyle\int_{-1}^1 (w^2 - 2w + 1)\,dw$

4. $\displaystyle\int_0^{\pi/2} \sin x\,dx$

5. $\displaystyle\int_0^1 e^x\,dx$

6. $\displaystyle\int_{e^3}^{e^9} \frac{1}{x}\,dx$

7. $\displaystyle\int_0^1 t^{90}\,dt$

8. $\displaystyle\int_1^8 \left(\frac{1}{\sqrt[3]{x}} + 5\sqrt[3]{x}\right) dx$

9. $\displaystyle\int_0^{\pi/4} \sec^2 x\,dx$

In problems 10–14, evaluate by expressing each sum as a definite integral.

10. $\displaystyle\lim_{n\to\infty} \frac{1}{n} \sum_{k=1}^n \frac{k}{n}$

11. $\displaystyle\lim_{n\to\infty} \frac{1}{n} \sum_{k=1}^n \frac{1}{1 + \dfrac{k}{n}}$

12. $\displaystyle\lim_{n\to\infty} \frac{1}{n} \sum_{k=1}^n e^{1 + \frac{k}{n}}$

13. $\displaystyle\lim_{n\to\infty} \frac{\pi/2}{n} \sum_{k=1}^n \sin\left(k \times \frac{\pi}{2n}\right)$

14. $\displaystyle\lim_{n\to\infty} \frac{3}{n} \sum_{k=1}^n \left(2 + \frac{3k}{n}\right)^3$

Solutions and Comments for Problem Set 6.2

1. $\left.\dfrac{x^4}{4}\right|_0^1 = 1/4.$

2. $\left.-\dfrac{1}{x}\right|_2^4 = -\dfrac{1}{4} - \left(-\dfrac{1}{2}\right) = 1/4.$

3. $\left.\left(\dfrac{w^3}{3} - w^2 + w\right)\right|_{-1}^1 = \left(\dfrac{1}{3} - 1 + 1\right) - (-\dfrac{1}{3} - 1 - 1) = 2\dfrac{2}{3}.$

4. $\left.-\cos x\right|_0^{\pi/2} = -0 - (-1) = 1.$

5. $\left.e^x\right|_0^1 = e^1 - e^0 = e - 1.$

6. $\left.\ln x\right|_{e^3}^{e^9} = \ln e^9 - \ln e^3 = 9 - 3 = 6.$

7. $\left.\dfrac{t^{91}}{91}\right|_0^1 = \dfrac{1}{91} - 0 = \dfrac{1}{91}.$

8. $\left.\left(\dfrac{3}{2}x^{2/3} + \dfrac{15}{4}x^{4/3}\right)\right|_1^8 = \left(\dfrac{3}{2} \times 4 + \dfrac{15}{4} \times 16\right) - \left(\dfrac{3}{2} + \dfrac{15}{4}\right) = 60\dfrac{3}{4}.$

9. $\left.\tan x\right|_0^{\pi/4} = 1 - 0 = 1.$

10. $b = 1, a = 0, f(x) = x$ in $\displaystyle\int_a^b f(x)\,dx.$

$\displaystyle\int_0^1 x\,dx = \left.\dfrac{x^2}{2}\right|_0^1 = \dfrac{1}{2}.$

11. $\displaystyle\int_0^1 \dfrac{dx}{1 + x} = \left.\ln(1 + x)\right|_0^1 = \ln 2.$

Alternate representation: $\displaystyle\int_1^2 \dfrac{dx}{x}.$

12. $\displaystyle\int_0^1 e^{1+x}\,dx = e^{1+x}\Big|_0^1 = e^2 - e.$

Alternate representation: $\displaystyle\int_1^2 e^x\,dx.$

$\boxed{\text{Dividing }\left[0,\dfrac{\pi}{2}\right]\text{ into }n\text{ equal subintervals.}}$

13. $\displaystyle\int_0^{\pi/2}\sin x = -\cos x\Big|_0^{\pi/2} = 1.$

14. $\displaystyle\int_2^5 x^3\,dx = \frac{x^4}{4}\Big|_2^5 = \frac{625}{4} - 4 = 152\tfrac{1}{4}.$

Alternate representation: $\displaystyle\int_0^3 (2+x)^3\,dx.$

Section 6.3: Basic Properties of the Definite Integral

The following represent some of the basic properties of definite integrals:

(1) $\displaystyle\int_a^b f(x)\,dx = -\int_b^a f(x)\,dx.$

(2) $\displaystyle\int_a^b kf(x)\,dx = k\int_a^b f(x)\,dx$ for any constant k.

(3) If $a < c < b$, then $\displaystyle\int_a^b f(x)\,dx = \int_a^c f(x)\,dx + \int_c^b f(x)\,dx.$

(4) $\displaystyle\int_a^b [f(x) + g(x)]\,dx = \int_a^b f(x)\,dx + \int_a^b g(x)\,dx.$

(5) $\displaystyle\int_a^a f(x)\,dx = 0.$

(6) If m and M are the least and greatest values of $f(x)$ in $[a, b]$, then $m(b - a) \le \displaystyle\int_a^b f(x)\,dx \le M(b - a).$

(7) *The Mean Value Theorem for Integrals:* There exists at least one $c \in [a, b]$ such that $\displaystyle\int_a^b f(x)\,dx = (b - a)f(c).$ The *mean value* of f with respect to x on $[a, b]$ is $\dfrac{\displaystyle\int_a^b f(x)\,dx}{b - a}.$

Authors' Note: Most of these properties are relatively obvious if we consider the appropriate definition and the geometry of the situation. For instance:

(1) By definition, $\displaystyle\int_a^b f(x)\,dx = \lim_{n\to\infty}\frac{b - a}{n}\times\sum_{j=1}^n f(c_j)$ and $\displaystyle\int_b^a f(x)\,dx = \lim_{n\to\infty}\frac{a - b}{n}\times\sum_{j=1}^n f(c_j).$

Since $\dfrac{b - a}{n} = -\dfrac{a - b}{n}$, it follows that $\displaystyle\int_a^b f(x)\,dx = -\int_b^a f(x)\,dx.$

(6) In the diagram, $\int_a^b f(x)\,dx$ is the shaded area and $f(a) = m$, $f(b) = M$. Also, $m(b - a)$ is the area of rectangle $ABCD$ and $M(b - a)$ is the area of rectangle $ABEF$. In this particular situation, it is obvious that

$$m(b - a) \le \int_a^b f(x)\,dx \le M(b - a).$$

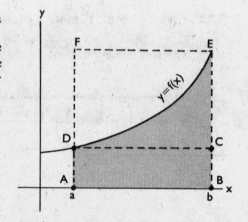

(7) Referencing the figure, there is a $c \in [a, b]$ such that rectangle $ABDE$, with height $= BD = f(c)$, has the same area as the shaded region.

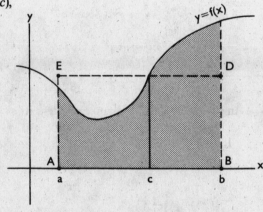

Q1. Evaluate $\int_0^4 |x - 2|\,dx$. _____

Answer: $\int_0^4 |x - 2|\,dx = \int_0^2 (2 - x)\,dx + \int_2^4 (x - 2)\,dx = \left(2x - \dfrac{x^2}{2}\right)\Big|_0^2 + \left(\dfrac{x^2}{2} - 2x\right)\Big|_2^4 =$
$[(4 - 2) - 0] + [(8 - 8) - (2 - 4)] = 2 + 2 = 4.$

Common error: $\int_0^4 |x - 2|\,dx = \int_0^4 (x - 2)\,dx = \left(\dfrac{x^2}{2} - 2x\right)\Big|_0^4 = 8 - 8 = 0.$

Authors' Note: $y = |x - 2|$ has no derivative at $x = 2$, but it is continuous there. Remember that differentiability is not a necessary condition for integration. Also, note the geometry of the situation. In Q1, we are merely asking for the total area of the shaded region; that is, the total area of the two triangles $= 4$ (sq. units).

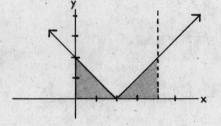

Q2. Consider $\int_1^3 x^3 \, dx$. Find the c guaranteed by the Mean Value Theorem for Integrals.

Answer: $\int_1^3 x^3 \, dx = \frac{x^4}{4}\Big|_1^3 = \frac{81}{4} - \frac{1}{4} = 20$. Then $20 = (3-1)c^3 \Rightarrow c^3 = 10 \Rightarrow c = \sqrt[3]{10}$.

Authors' Note: In Q2, the mean value of $f(x) = x^3$ on $[1,3]$ is $\frac{20}{3-1} = 10$. This represents the "average" of the y values as x ranges from 1 to 3.

Problem Set 6.3

In problems 1–8, evaluate the integrals.

1. $\int_0^1 (x^2 + e^x) \, dx$

2. $\int_2^4 \left(\frac{1}{z^3} + \frac{1}{z^2} \right) dz$

3. $\int_0^{\sqrt{3}/2} \frac{dt}{\sqrt{1-t^2}}$

4. $\int_0^2 xe^x \, dx$

5. $\int_{-3}^2 (4x^{32} + 7) \, dx + \int_2^{-3} (4x^{32} + 7) \, dx$

6. $\int_{-2}^4 |x| \, dx$

7. $\int_5^5 e^x \ln x \, dx$

8. $\int_{-1}^4 f(x) \, dx$ where $f(x) = \begin{cases} x^2 & \text{if } x \geq 1 \\ 4x - 3 & \text{if } x < 1 \end{cases}$

9. If $f(x) = \begin{cases} x + 4 & \text{if } x \geq 0 \\ x - 4 & \text{if } x < 0 \end{cases}$, explain why $\int_{-2}^2 f(x) \, dx$ is undefined.

10. If f is a function continuous on $[5,9]$ and if $12 \leq f(x) \leq 16$ for all $x \in [5,9]$, explain why $48 \leq \int_5^9 f(x) \, dx \leq 64$.

11. Show that $\int_0^3 \frac{dx}{(x^3 + 9)^{1/2}}$ is between $\frac{1}{2}$ and 1.

12. Find all of the c's guaranteed by the Mean Value Theorem for Integrals.

    (a) $\int_1^2 x^2 \, dx$

    (b) $\int_{-2}^2 (x^2 + 1) \, dx$

    (c) $\int_0^1 e^{3x} \, dx$

13. Find the mean value of f with respect to x on the designated interval.

    (a) $f(x) = x^3 + 2$ on $[0, 3]$
    (b) $f(x) = \sin x$ on $[0, \pi]$
    (c) $f(x) = e^{6x}$ on $[-1, 1]$

Solutions and Comments for Problem Set 6.3

1. $\left(\dfrac{x^3}{3} + e^x\right)\Big|_0^1 = -\dfrac{2}{3} + e.$

2. $-\left(\dfrac{1}{2z^2} + \dfrac{1}{z}\right)\Big|_2^4 = \dfrac{11}{32}.$

3. $\sin^{-1} t \Big|_0^{\sqrt{3}/2} = \dfrac{\pi}{3} - 0 = \pi/3.$

4. $(xe^x - e^x)\Big|_0^2 = e^2 + 1.$ (Integration by parts)

5. 0. Make use of property (1) in the reading section.

6. $\displaystyle\int_{-2}^0 (-x)\,dx + \int_0^4 x\,dx = 10.$

7. 0. Use property (5) in the reading section.

8. $\displaystyle\int_{-1}^1 (4x - 3)\,dx + \int_1^4 x^2\,dx = 15.$
   The function is continuous at $x = 1$.

9. f is not continuous throughout $[-2, 2]$ since it has a point of discontinuity at $x = 0$.

> The definition in Section 6.2 requires that f be continuous over the interval of integration.

10. Property (6) in the reading section justifies the stated inequality.
   $$12(9 - 5) \le \int_5^9 f(x)\,dx \le 16(9 - 5).$$

11. On $[0, 3]$, the minimum value of $f(x) = \dfrac{1}{(x^3 + 9)^{1/2}}$ is $f(3) = \dfrac{1}{6}$ and the maximum value is $f(0) = \frac{1}{3}$. Property (6) of the reading sections

then justifies the statement
$$\frac{1}{6}(3 - 0) \le \int_0^3 \frac{dx}{(x^3 + 9)^{1/2}} \le \frac{1}{3}(3 - 0).$$

12. (a) $\dfrac{x^3}{3}\Big|_1^2 = \dfrac{7}{3} = (2 - 1)c^2 \Rightarrow c = \sqrt{\dfrac{7}{3}}$ on $[1, 2]$.

   (b) $\left(\dfrac{x^3}{3} + x\right)\Big|_{-2}^2 = \dfrac{28}{3} = [2 - (-2)](c^2 + 1)$
   $$\Rightarrow c^2 = \frac{4}{3} \Rightarrow c = \pm\sqrt{\frac{4}{3}} \text{ on } [-2, 2].$$

   (c) $\dfrac{1}{3}e^{3x}\Big|_0^1 = \dfrac{1}{3}(e^3 - 1) = (1 - 0)e^{3c} \Rightarrow$
   $$3c = \ln\frac{e^3 - 1}{3} \Rightarrow c = \frac{1}{3}\ln\frac{e^3 - 1}{3} =$$
   $$\ln\sqrt[3]{\frac{e^3 - 1}{3}}.$$

> In (b), if one considers the graph of $y = x^2 + 1$, it should not be surprising that there are two values of c in the interval $[-2, 2]$.
>
>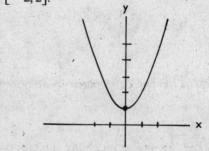

13. (a) $8\frac{3}{4}$;
   (b) $\frac{2}{\pi}$;
   (c) $\frac{1}{12}(e^6 - e^{-6})$.

> Just calculate $\dfrac{\displaystyle\int_a^b f(x)\,dx}{b - a}.$

Section 6.4: Variables as Limits of Integration

> If a is a constant, then
> $$\frac{d}{dx}\int_a^x f(x)\,dx = f(x).$$

Authors' Note: This is relatively easy to establish. If $\int f(x)\,dx = F(x) + C$, then $\int_a^x f(x)\,dx = (F(x) + C)\Big|_a^x =$

$(F(x) + C) - (F(a) + C) = F(x) - F(a)$. Hence, $\dfrac{d}{dx}\displaystyle\int_a^x f(x)\,dx = F'(x) - F'(a) = f(x) - 0 =$

$f(x)$. We should also note that we can use any variable as the variable of integration since the definite integral depends only on the limits of integration. For instance,

$$\frac{d}{dx}\int_a^x f(x)\,dx = \frac{d}{dx}\int_a^x f(t)\,dt = \frac{d}{dx}\int_a^x f(w)\,dw.$$

If the upper and lower integration limits are variables, then

$$\frac{d}{dx}\int_{h(x)}^{g(x)} f(t)\,dt = f(g(x))g'(x) - f(h(x))h'(x).$$

Q1. Find: (a) $\dfrac{d}{dx}\displaystyle\int_5^x \frac{e^t}{t}\,dt$ _____ (b) $\dfrac{d}{dx}\displaystyle\int_x^{x^3} \sin t\,dt$ _____

Answers: (a) $\dfrac{e^x}{x}$. (b) $\sin x^3 \dfrac{d}{dx}(x^3) - \sin x\,\dfrac{d}{dx}(x) = 3x^2 \sin x^3 - \sin x.$

If $a \le x \le b$, then we can define a function

$$G(x) = \int_a^x f(t)\,dt.$$

The values for $G(x)$ are determined by the points in $[a, b]$. The function G is referred to as an indefinite integral of f.

Authors' Note: We say *an* indefinite integral rather than *the* indefinite integral because G depends on the lower limit a. A different lower limit, say c, will determine another indefinite integral of f.

Q2. Find $G(x)$ if $G(x) = \displaystyle\int_2^x (1 - t^2)\,dt.$ _____

Answer: $G(x) = \left(t - \dfrac{t^3}{3}\right)\Big|_2^x = x - \dfrac{x^3}{3} + \dfrac{2}{3}.$

Functions may be determined if both upper and lower limits of integration are variables.

Q3. Find $H(x)$ if $H(x) = \displaystyle\int_{-x}^{x^2} e^{2w}\,dw.$ _____

Answer: $H(x) = \dfrac{1}{2}e^{2w}\Big|_{-x}^{x^2} = \dfrac{1}{2}e^{2x^2} - \dfrac{1}{2}e^{-2x} = \dfrac{1}{2}\left(e^{2x^2} - \dfrac{1}{e^{2x}}\right).$

Problem Set 6.4

In problems 1–8, evaluate the expression.

1. $\dfrac{d}{dx}\displaystyle\int_3^x t^3 e^{4t}\,dt$

2. $\dfrac{d}{dx}\displaystyle\int_{-2}^x \sqrt{1 + \cos^2 x}\,dx$

3. $\dfrac{d}{dx}\displaystyle\int_{x^2}^{x^3+1} \dfrac{\cos w}{w}\,dw$

4. $\dfrac{d}{dx}\displaystyle\int_x^3 t^3 \sin t\,dt$

5. $\dfrac{d}{dx}\displaystyle\int_{\cos x}^{\sin x} (t^4 + 2t)\,dt$

6. $\displaystyle\int_0^x (t + 3t^4)\,dt$

7. $\displaystyle\int_{\pi/2}^x \left(\sin 2w + \cos\dfrac{w}{2}\right)dw$

8. $\displaystyle\int_1^{x^2} \left(\dfrac{1}{2} - \dfrac{1}{x}\right)dx$

9. If $F_1(x) = \displaystyle\int_a^x f(t)\,dt$ and $F_2(x) = \displaystyle\int_b^x f(t)\,dt$, show that $F_1(x) - F_2(x)$ is independent of x. That is, show that this difference is a constant.

Solutions and Comments for Problem Set 6.4

1. $x^3 e^{4x}$.

2. $\sqrt{1 + \cos^2 x}$.

Problems #1 and #2 represent a direct application of the first formula in the reading section.

3. $\dfrac{\cos(x^3 + 1)}{x^3 + 1}(3x^2) - \dfrac{\cos(x^2)}{x^2}(2x)$.

A direct application of the second formula in the reading section.

4. $\displaystyle\int_x^3 t^3 \sin t\,dt = -\int_3^x t^3 \sin t\,dt = -x^3 \sin x$.

5. $(\sin^4 x + 2\sin x)(\cos x) - $
   $(\cos^4 x + 2\cos x)(-\sin x) =$
   $\sin^4 x \cos x + \sin x \cos^4 x + 4\sin x \cos x$.

Ditto comment for problem #3.

6. $\left(\dfrac{t^2}{2} + \dfrac{3t^5}{5}\right)\Big|_0^x = \dfrac{x^2}{2} + \dfrac{3x^5}{5}$.

7. $\left(-\dfrac{1}{2}\cos 2w + 2\sin\dfrac{w}{2}\right)\Big|_{\pi/2}^x =$
   $-\dfrac{1}{2}\cos 2x + 2\sin\dfrac{x}{2} - \dfrac{1}{2} - \sqrt{2}$.

8. $\left(\dfrac{x}{2} - \ln|x|\right)\Big|_1^{x^2} = \dfrac{x^2}{2} - \ln x^2 - \dfrac{1}{2}$.

Note that x cannot be zero.

9. $\displaystyle\int_a^x f(t)\,dt - \int_b^x f(t)\,dt = \int_a^b f(t)\,dt$.

$\displaystyle\int_a^b f(t)\,dt$ is a constant (and hence independent of x).

Section 6.5: Area Between Curves

Previous sections have related the definite integral $\int_a^b f(x)\,dx$ to the concept of area.

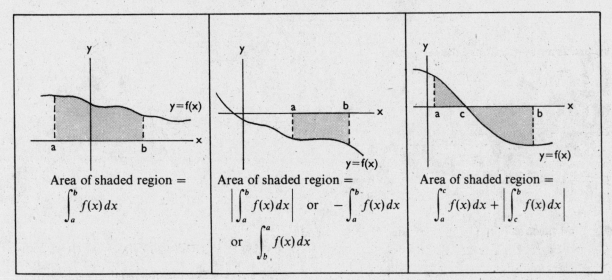

Area of shaded region =

$$\int_a^b f(x)\,dx$$

Area of shaded region =

$$\left|\int_a^b f(x)\,dx\right| \quad \text{or} \quad -\int_a^b f(x)\,dx$$

$$\text{or} \quad \int_b^a f(x)\,dx$$

Area of shaded region =

$$\int_a^c f(x)\,dx + \left|\int_c^b f(x)\,dx\right|$$

Authors' Note: The reader is again reminded that $\int_a^b f(x)\,dx = \lim\limits_{n\to\infty} \dfrac{b-a}{n} \sum\limits_{k=1}^{n} f(c_k)$, where c_k is any point in the k-th subinterval $[x_{k-1}, x_k]$. If $f(x) < 0$ on an interval $[p, q]$, then $\int_p^q f(x)\,dx < 0$ since $f(c_k) < 0$ for $c_k \in [p, q]$. Hence area would somehow have to be expressed as the opposite of the value $\int_p^q f(x)\,dx$.

Q1. The graph is $y = \sin x$. Find the total area of the shaded region. _____

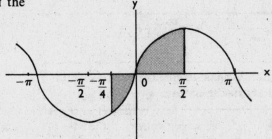

Answer: $\left|\displaystyle\int_{-\pi/4}^{0} \sin x\,dx\right| + \displaystyle\int_{0}^{\pi/2} \sin x\,dx = \left|-1 + \dfrac{\sqrt{2}}{2}\right| + 1 = 1 - \dfrac{\sqrt{2}}{2} + 1 = 2 - \dfrac{\sqrt{2}}{2} = \dfrac{4 - \sqrt{2}}{2}$

(sq. units).

Common error: In Q1, total area of shaded region $= \displaystyle\int_{-\pi/4}^{\pi/2} \sin x \, dx =$
$-\cos x \Big|_{-\pi/4}^{\pi/2} = \dfrac{\sqrt{2}}{2}$ (sq. units).

If f and g are continuous functions such that $f(x) \geq g(x)$ on $[a, b]$, then the area between the curves is $\displaystyle\int_a^b [f(x) - g(x)] \, dx.$

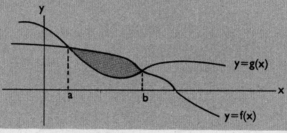

Q2. Find the area of the region bounded by $y = x^2$ and $y = 2x$. _____

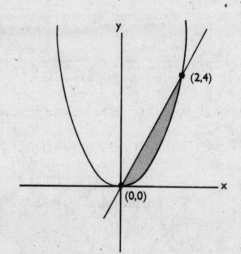

Answer: The curves intersect at $(0, 0)$ and $(2, 4)$. The desired area is
$$\int_0^2 (2x - x^2) \, dx = \left(x^2 - \frac{x^3}{3}\right)\Big|_0^2$$
$$= 4 - \tfrac{8}{3}$$
$$= 1\tfrac{1}{3} \text{ (sq. units)}.$$

Authors' Note: If we think of adding up the areas of a lot of "little" rectangles to get the area between two curves, the base of a typical rectangle is $\Delta x = \dfrac{b - a}{n}$ and the height is $f(c_j) - g(c_j)$. Hence the expression $\displaystyle\int_a^b [f(x) - g(x)] \, dx$ should not seem "foreign" to the reader.

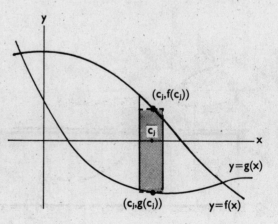

In some instances, it may be beneficial to integrate with respect to y. For instance, if we want the area bounded by $y^2 = 4x$ and $y = 2x - 4$, then solving for x, we have $x = \dfrac{y^2}{4}$ and $x = \dfrac{y}{2} + 2$. The desired area is

$$\int_{-2}^{4} \left(\frac{y}{2} + 2 - \frac{y^2}{4} \right) dy = \left(\frac{y^2}{4} + 2y - \frac{y^3}{12} \right) \Bigg|_{-2}^{4} = 9 \text{ (sq. units)}.$$

A typical rectangle with base Δy is shown in the figure.

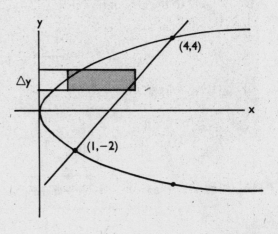

Problem Set 6.5

1. Express the area of the shaded regions as a sum of definite integrals.

(a)

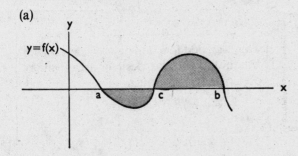

(b)

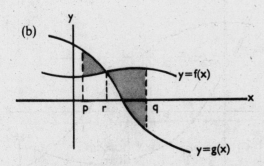

2. Find the area bounded by $y = x^2$ and $y = x$.

3. Find the area bounded by $y^2 = 4x$ and $x^2 = 4y$.

4. The following figure shows the graphs of $y = x^2$, $y = 4x$, and $x = 1$. Find the area of the shaded region.

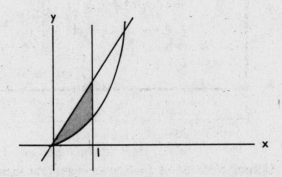

5. Find the first quadrant area bounded by $y = \sin x$, $y = \cos x$, and the y-axis.

6. Find the area bounded by $y = x^3 - 4x + 10$, $x = 1$, $x = 2$, and the x-axis.

7. Find the area bounded by $y = e^{-x}$, the x-axis, $x = 1$, and $x = 4$.

8. Find the area bounded by $y^2 = 4 + x$ and $x + 2y = 4$.

9. Find the area enclosed by the x-axis and the loops of the curve $y = 4x - x^3$.

10. The following graph is a portion of $y = \cos x$. Find the value of c so that the area of region A is equal to the area of region B.

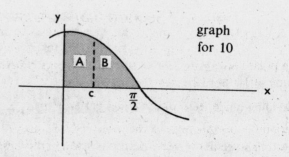

graph for 10

Solutions and Comments for Problem Set 6.5

1. (a) $\left| \int_a^c f(x)\,dx \right| + \int_c^b f(x)\,dx;$

   (b) $\int_p^r [g(x) - f(x)]\,dx + \int_r^q [f(x) - g(x)]\,dx.$

   > There are other ways to represent the areas.

2. $\int_0^1 (x - x^2)\,dx = \frac{1}{6}$ (sq. units).

3. $\int_0^4 \left(2x^{1/2} - \frac{x^2}{4} \right) dx = 5\frac{1}{3}$ (sq. units).

4. $\int_0^1 (4x - x^2)\,dx = 1\frac{2}{3}$ (sq. units).

5. $\int_0^{\pi/4} (\cos x - \sin x)\,dx = (\sin x + \cos x)\Big|_0^{\pi/4} = \sqrt{2} - 1$ (sq. units).

6. $\int_1^2 (x^3 - 4x + 10)\,dx = 7\frac{3}{4}$ (sq. units).

   > Note that the function is above the x-axis in the interval $[1,2]$. It is important to check this.

7. $\int_1^4 e^{-x}\,dx = -e^{-x}\Big|_1^4 = \frac{1}{e} - \frac{1}{e^4} = \frac{e^3 - 1}{e^4}$ (sq. units).

8. $\int_{-4}^2 [(4 - 2y) - (y^2 - 4)]\,dy = 36$ (sq. units).

> It is definitely easier to integrate with respect to y in this problem.

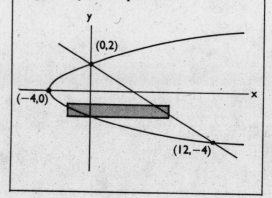

9. $\left| \int_{-2}^0 (4x - x^3)\,dx \right| + \int_0^2 (4x - x^3)\,dx = 8$ (sq. units).

> This is an odd function and the alert reader will note that the desired area can be obtained by calculating.
>
> $2\int_0^2 (4x - x^3)\,dx$. Note that
>
> $\int_{-2}^2 (4x - x^3)\,dx = 0.$

10. $\int_0^c \cos x\,dx = \int_c^{\pi/2} \cos x\,dx \Rightarrow \sin x\Big|_0^c = \sin x\Big|_c^{\pi/2}$

$\Rightarrow \sin c = 1 - \sin c \Rightarrow 2\sin c = 1 \Rightarrow$
$\sin c = \frac{1}{2} \Rightarrow c = \frac{\pi}{6}.$

*Section 6.5A: Areas with Polar Coordinates

In polar coordinates, if $r = f(\theta)$ we can approximate the area AOB bordered by $\theta = \alpha$, $\theta = \beta$, and $r = f(\theta)$ by dividing angle AOB into n equal parts $\Delta\theta = \dfrac{\beta - \alpha}{n}$ and summing up the areas of the n sectors. The area of the sector shown in the figure is approximately $\frac{1}{2}r_i^2\Delta\theta$, which is the area of triangle POQ. Hence, the area of the region AOB is approximately

$$\frac{1}{2}\sum_{i=1}^{n} r_i^2 \Delta\theta = \frac{1}{2}\sum_{i=1}^{n} f^2(\theta_i)\Delta\theta.$$

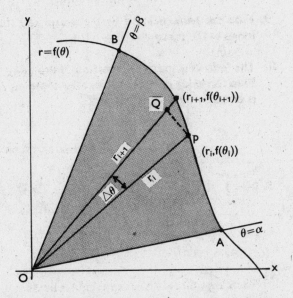

The actual area of the region bounded by $r = f(\theta)$, $\theta = \alpha$, and $\theta = \beta$ is

$$\lim_{n\to\infty}\frac{1}{2}\sum_{i=1}^{n} f^2(\theta_i)\Delta\theta = \frac{1}{2}\int_\alpha^\beta f^2(\theta)\,d\theta = \frac{1}{2}\int_\alpha^\beta r^2\,d\theta.$$

Authors' Notes: The trig identities $\sin^2\theta = \frac{1}{2}(1 - \cos 2\theta)$ and $\cos^2\theta = \frac{1}{2}(1 + \cos 2\theta)$ are derived easily from the well-known double angle formulas and often come in handy when working with polar coordinates. Also, the reduction formulas (2) and (3) in Section 5.5A are often useful.

Example: To find the area of the region enclosed by the cardioid $r = 1 + \sin\theta$, we can note that the graph is symmetric with respect to $\theta = \frac{\pi}{2}$. Hence we can integrate from $-\frac{\pi}{2}$ to $\frac{\pi}{2}$ and double the value of the integral. The desired area is

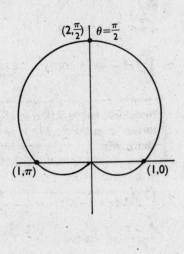

$$2 \times \frac{1}{2}\int_{-\pi/2}^{\pi/2} (1 + \sin\theta)^2\,d\theta = \int_{-\pi/2}^{\pi/2} (1 + 2\sin\theta + \sin^2\theta)\,d\theta$$

$$= \int_{-\pi/2}^{\pi/2} \left(1 + 2\sin\theta + \frac{1}{2}(1 - \cos 2\theta)\right)d\theta$$

$$= \int_{-\pi/2}^{\pi/2} \left(\frac{3}{2} + 2\sin\theta - \frac{1}{2}\cos 2\theta\right)d\theta$$

$$= \left(\frac{3}{2}\theta - 2\cos\theta - \frac{1}{4}\sin 2\theta\right)\Bigg|_{-\pi/2}^{\pi/2}$$

$$= \left(\frac{3\pi}{4} - 0 - 0\right) - \left(-\frac{3\pi}{4} - 0 - 0\right)$$

$$= \frac{3\pi}{2} \text{ (sq. units)}.$$

Problem Set 6.5A

Find the areas bounded by the following:

1. $r = 5$, $\theta = 0$, $\theta = \pi/2$

2. $r = 3 \sin \theta$

3. $r = \cos 3\theta$

4. $r = \cos \theta + \sin \theta$

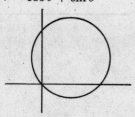

5. One loop of $r = 4 \cos 4\theta$

6. One loop of $r^2 = a^2 \sin 2\theta$

7. The cardioid $r = a(1 - \cos \theta)$

8. Find the area inside the circle $r = 5 \sin \theta$ and outside the limacon $r = 2 + \sin \theta$

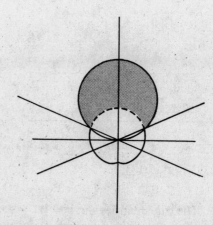

Solutions and Comments for Problem Set 6.5A

1. $\dfrac{1}{2} \displaystyle\int_0^{\pi/2} 5^2 \, d\theta = \dfrac{1}{2} \times 25\theta \Big|_0^{\pi/2} = \dfrac{25\pi}{4}$ (sq. units).

> This is just one-fourth of the area of a circle with radius = 5.

> The graph is a circle with center at $(0, \frac{3}{2})$ in rectangular coordinates and radius = $\frac{3}{2}$.

2. $\dfrac{1}{2} \displaystyle\int_0^{\pi} 9 \sin^2 \theta \, d\theta = \dfrac{9}{2} \int_0^{\pi} \dfrac{1}{2}(1 - \cos 2\theta) \, d\theta =$

$\dfrac{9}{4}\left[\theta - \dfrac{1}{2}\sin 2\theta\right]\Big|_0^{\pi} = \dfrac{9\pi}{4}$ (sq. units).

3. $\dfrac{1}{2} \times 6 \displaystyle\int_0^{\pi/6} \cos^2 3\theta \, d\theta = 3 \int_0^{\pi/6} \dfrac{1}{2}(1 + \cos 6\theta) \, d\theta$

$= \dfrac{3}{2}\left[\theta + \dfrac{1}{6}\sin 6\theta\right]\Big|_0^{\pi/6} = \dfrac{\pi}{4}$ (sq. units).

There are three loops (see graph). Take 6 times the area of one-half of one loop. One-half of a loop is generated as θ goes from 0 to $\pi/6$.

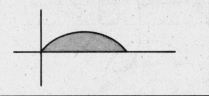

One-half of one loop is generated as θ increases from 0 to $\frac{\pi}{8}$.

4. $\dfrac{1}{2}\displaystyle\int_0^\pi (\cos\theta + \sin\theta)^2\, d\theta =$

$\dfrac{1}{2}\displaystyle\int_0^\pi (1 + 2\sin\theta\cos\theta)\, d\theta = \dfrac{1}{2}\displaystyle\int_0^\pi (1 + \sin 2\theta)\, d\theta$

$= \dfrac{1}{2} \times \left[\theta - \dfrac{1}{2}\cos 2\theta \right] \Bigg|_0^\pi = \dfrac{\pi}{2}$ (sq. units).

6. $\dfrac{1}{2}\displaystyle\int_0^{\pi/2} a^2 \sin 2\theta\, d\theta = -\dfrac{a^2}{4} \times \cos 2\theta \Bigg|_0^{\pi/2} =$
$\frac{1}{2}a^2$ (sq. units).

$r = 0$ when $\theta = 0$, and r becomes 0 again when $\theta = \frac{\pi}{2}$; this completes a loop.

Rectangular equation is
$(x - \frac{1}{2})^2 + (y - \frac{1}{2})^2 = \frac{1}{2}$.

7. $2 \times \dfrac{1}{2}\displaystyle\int_0^\pi [a(1 - \cos\theta)]^2\, d\theta = \dfrac{3}{2}\pi a^2$ (sq. units).

This problem is similar to the example in the reading section.

5. $\dfrac{1}{2} \times 2 \displaystyle\int_0^{\pi/8} 16\cos^2 4\theta\, d\theta =$

$16 \displaystyle\int_0^{\pi/8} \dfrac{1}{2}(1 + \cos 8\theta)\, d\theta = 8\left[\theta + \dfrac{1}{8}\sin 8\theta \right] \Bigg|_0^{\pi/8} =$
π (sq. units).

8. $2\left[\dfrac{1}{2}\displaystyle\int_{\pi/6}^{\pi/2} (5\sin\theta)^2\, d\theta - \dfrac{1}{2}\displaystyle\int_{\pi/6}^{\pi/2} (2 + \sin\theta)^2\, d\theta \right] =$
$\dfrac{8\pi}{3} + \sqrt{3}$ (sq. units).

The curves intersect when $2 + \sin\theta = 5\sin\theta$, that is, when $\theta = \frac{\pi}{6}$ and $\theta = \frac{5\pi}{6}$. Calculate the area of the circle between $\frac{\pi}{6}$ and $\frac{5\pi}{6}$ and subtract from it the area of the limacon between $\frac{\pi}{6}$ and $\frac{5\pi}{6}$.

Section 6.6: Volumes of Solids of Revolution: Disk and Washer Methods

If f is continuous on $[a, b]$, consider the region bounded by the graph of $y = f(x)$, the x-axis, and the lines $x = a$ and $x = b$. If this region is revolved about the x-axis, a *solid of revolution* is generated. If we divide $[a, b]$ into n subintervals of length $\Delta x = \dfrac{b - a}{n}$, we can approximate the volume of this solid by adding up the volumes of n right circular cylinders that are formed when rectangular regions (see figure) are rotated about the x-axis. The volume of the cylinder shown in the figure is $\pi[f(x_i)]^2 \Delta x$ and the volume of the solid (top of page 168) formed by rotating the entire region is

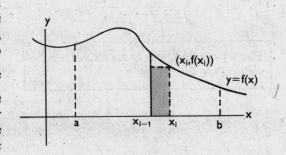

$$\sum_{i=1}^{n} \pi [f(x_i)]^2 \Delta x = \pi \left(\frac{b-a}{n} \right) \sum_{i=1}^{n} [f(x_i)]^2.$$

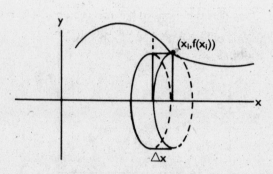

The actual volume is

$$\lim_{n \to \infty} \pi \sum_{i=1}^{n} [f(x_i)]^2 \Delta x = \pi \int_a^b [f(x)]^2 \, dx = \pi \int_a^b y^2 \, dx.$$

The method of obtaining volumes by means of right circular cylinders (or disks) is called the *disk method*.

Q1. The region bounded by $y = x^2$, $x = 1$, $x = 3$, and the x-axis is revolved about the x-axis. Find the volume of the solid generated. _____

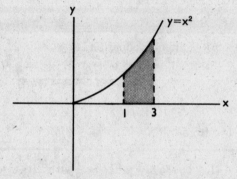

Answer: $\pi \displaystyle\int_1^3 y^2 \, dx = \pi \int_1^3 x^4 \, dx = \pi \dfrac{x^5}{5} \Big|_1^3$

$\qquad = \dfrac{242\pi}{5}$ (cu. units).

 Common error: In Q1, volume $= \displaystyle\int_1^3 x^4 \, dx = \dfrac{242}{5}$ (cu. units).

Authors' Note: The disk method can also be used to find the volumes of solids formed by revolving a region about the y-axis or about a line parallel to the x- or y-axis. The "key" is to think of a typical rectangular region that is revolving about an axis to form one of the right circular cylinders that "make up" the solid, as demonstrated in the following example. Just remember that the volume of a right circular cylinder with base radius r and altitude h is $\pi r^2 h$. Also, note that the volume of a ring (or washer) between two concentric right circular cylinders is $\pi R^2 h - \pi r^2 h = \pi h(R^2 - r^2)$. Note *very carefully* that $\pi h(R^2 - r^2)$ is *not* the same as $\pi h(R - r)^2$.

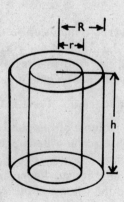

Example: Consider the first quadrant region *OAB* bounded by $y^2 = x$, the *x*-axis, and the line $x = 4$.

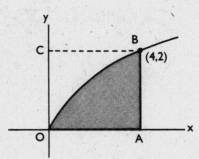

Rotate *OAB* about the *x*-axis. Typical cylinder has radius $y - 0 = y$ and altitude Δx.

$$\text{Volume} = \pi \int_0^4 y^2 \, dx = \pi \int_0^4 x \, dx = \pi \frac{x^2}{2} \Big|_0^4 = 8\pi \text{ (cu. units)}.$$

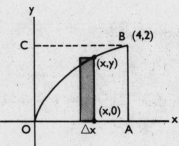

Rotate *OAB* about the *y*-axis.
We need (volume of rotated region *OABC*) −
(volume of rotated region *OBC*).

$$\text{Volume} = \pi \int_0^2 4^2 \, dy - \pi \int_0^2 x^2 \, dy = \pi \int_0^2 (16 - x^2) \, dy$$

$$= \int_0^2 (16 - y^4) \, dy = \frac{128\pi}{5} \text{ (cu. units)}.$$

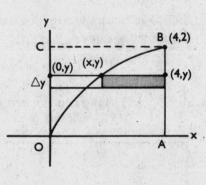

Rotate *OAB* about $y = -1$.
We need (volume of rotated region *FEBO*) −
(volume of rotated region *FEAO*).

$$\text{Volume} = \pi \int_0^4 [y - (-1)]^2 \, dx - \pi \int_0^4 1^2 \, dx$$

$$= \pi \int_0^4 (x^{1/2} + 1)^2 \, dx - \pi \int_0^4 1 \, dx$$

$$= \pi \int_0^4 (x + 2x^{1/2}) \, dx = \frac{56\pi}{3} \text{ (cu. unit)}.$$

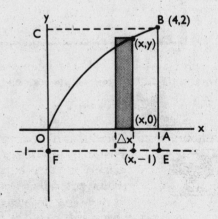

Rotate *OAB* about $x = 6$.
We need (volume of rotated region *OGHB*) −
(volume of rotated region *AGHB*).

$$\text{Volume} = \pi \int_0^2 (6-x)^2\, dy - \pi \int_0^2 2^2\, dy$$

$$= \pi \int_0^2 (6-y^2)^2\, dy - \pi \int_0^2 4\, dy$$

$$= \pi \int_0^2 (20 - 12y^2 + y^4)\, dy = \frac{72\pi}{5} \text{ (cu. units)}.$$

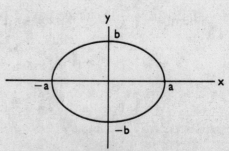

The method of calculating volume as the difference of the volumes of two concentric cylinders or disks (as we did in the last three parts of the example) is known as the *washer method* (sometimes called the *ring method*).

 Common error (Reference: first part of example): $\pi \int_0^4 y^2\, dx = \pi \dfrac{y^3}{3}\Big|_0^4$.

 Common error (Reference: second part of example): Volume about y-axis is $\pi \int_0^2 (4-x)^2\, dy$.

 Authors' Note: The two common errors above should alert the reader
          (1) to be aware of the variable of integration;
          (2) to remember that $(4-x)^2 \neq 4^2 - x^2$. Note that $4-x$ *is not* the radius of any rotating right circular cylinder.

Problem Set 6.6

In problems 1–5, find the volumes obtained by rotating about the x-axis the regions bounded by the indicated curves.

1. $y = x^2$, $x = -1$, $x = 4$, $y = 0$

2. $y^2 = 4x$, $x = 2$

3. $y = \sin x$, $x = 0$, $x = \pi$, $y = 0$

4. $y^4 = x$, $x = 4$

5. $y = \sqrt{r^2 - x^2}$, $y = 0$

6. Find the volume of the solid generated by rotating the region bounded by $y = x^2$, $y = 2$, and $x = 0$ about the y-axis.

7. Consider the ellipse $\dfrac{x^2}{a^2} + \dfrac{y^2}{b^2} = 1$.

Find the volume obtained
(a) by rotating the first and second quadrant regions about the x-axis
(b) by rotating the first and fourth quadrant regions about the y-axis

8. Consider the region enclosed by $y = x^2$ and $y = x$. Find the volume of the solid generated if this region is
   (a) rotated about the x-axis
   (b) rotated about the y-axis

9. Consider the region bounded by the y-axis, $y = 4$, and $y = x^2$. Find the volume obtained if

this region is rotated about
   (a) the y-axis
   (b) the x-axis
   (c) the line $y = 4$
   (d) the line $x = 2$
   (e) the line $x = 4$
   (f) the line $y = 6$

Solutions and Comments for Problem Set 6.6

1. $\pi \int_{-1}^{4} x^4 \, dx = 205\pi$ (cu. units).

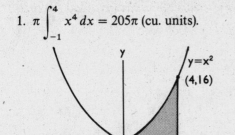

2. $\pi \int_{0}^{2} 4x \, dx = 8\pi$ (cu. units).

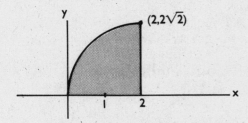

3. $\pi \int_{0}^{\pi} \sin^2 x \, dx = \pi \int_{0}^{\pi} \frac{1}{2}(1 - \cos 2x) \, dx$
   $= \frac{\pi}{2}\left(x - \frac{1}{2}\sin 2x\right)\Big|_{0}^{\pi} = \frac{\pi^2}{2}$ (cu. units).

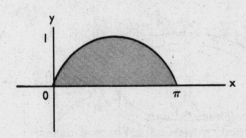

4. $\pi \int_{0}^{4} x^{1/2} \, dx = \frac{16\pi}{3}$ (cu. units).

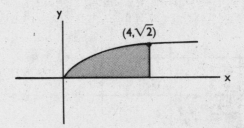

5. $\pi \int_{-r}^{r} (r^2 - x^2) \, dx = 2\pi \int_{0}^{r} (r^2 - x^2) \, dx =$
   $2\pi\left(r^2 x - \frac{x^3}{3}\right)\Big|_{0}^{r} = \frac{4}{3}\pi r^3$ (cu. units).

> Note that a sphere is being generated by rotating a semicircle about the x-axis. We have just derived the formula for the volume of a sphere in terms of its radius.

6. $\pi \int_{0}^{2} x^2 \, dy = \pi \int_{0}^{2} y \, dy = 2\pi$ (cu. units).

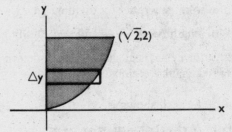

7. (a) $2\pi \int_{0}^{a} \left(b^2 - \frac{b^2}{a^2}x^2\right) dx = \frac{4}{3}\pi ab^2$ (cu. units).

   (b) $2\pi \int_{0}^{b} \left(a^2 - \frac{a^2}{b^2}y^2\right) dy = \frac{4}{3}\pi a^2 b$ (cu. units).

> (a) Solid is called a prolate spheroid.
> (b) Solid is called a oblate spheroid.

8.

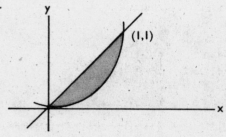

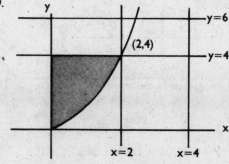

(a) $\pi \int_0^1 (x^2 - x^4)\,dx = \dfrac{2\pi}{15}$ (cu. units);

(b) $\pi \int_0^1 (y - y^2)\,dy = \dfrac{\pi}{6}$ (cu. units).

9.

(a) $\pi \int_0^4 y\,dy = 8\pi$ (cu. units).

(b) $\pi \int_0^2 (4^2 - x^4)\,dx = \dfrac{128\pi}{5}$ (cu. units).

(c) $\pi \int_0^2 (x^2 - 4)^2\,dx = \dfrac{256\pi}{15}$ (cu. units).

(d) $\pi \int_0^4 \left[(-2)^2 - (\sqrt{y} - 2)^2\right]\,dy$

$= \dfrac{40\pi}{3}$ (cu. units).

(e) $\pi \int_0^4 \left[(-4)^2 - (\sqrt{y} - 4)^2\right]\,dy$

$= \dfrac{104\pi}{3}$ (cu. units).

(f) $\pi \int_0^2 \left[(x^2 - 6)^2 - (4 - 6)^2\right]\,dx$

$= \dfrac{192\pi}{5}$ (cu. units).

Section 6.7: Volumes of Solids of Revolution: Shell Method

Consider a function $y = f(x)$ continuous on $[a, b]$. We will rotate this region about the y-axis and calculate the volume of the solid generated by dividing $[a, b]$ into n subintervals of length $\Delta x = \dfrac{b - a}{n}$ and then "adding up" the volumes of the cylindrical shells that are determined by revolving rectangular regions.

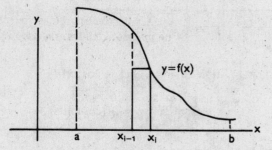

The volume of the shell shown is

$$\pi x_i^2 f(x_i) - \pi x_{i-1}^2 f(x_i) = \pi f(x_i)(x_i^2 - x_{i-1}^2) =$$

$$\pi f(x_i)(x_i + x_{i-1})(x_i - x_{i-1}) = 2\pi f(x_i)\left(\dfrac{x_i + x_{i-1}}{2}\right)\Delta x.$$

Now $\dfrac{x_i - x_{i-1}}{2}$, which we will call x_i^*, is the midpoint of the interval $[x_{i-1}, x_i]$. Hence the approximate volume of the solid generated by the rotating region is $2\pi \sum\limits_{i=1}^{n} x_i^* f(x_i)\Delta x$.

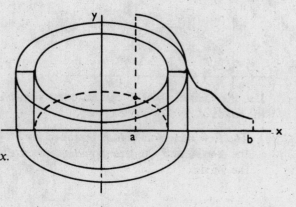

The actual volume, calculated by the *shell method*, is

$$\lim_{n \to \infty} 2\pi \sum_{i=1}^{n} x_i^* f(x_i) \Delta x = 2\pi \int_a^b x f(x)\, dx.$$

A similar formula holds if $x = g(y)$ and the rotation is about the x-axis. Referring to the figure, the volume obtained by rotating the shaded region about the x-axis is $2\pi \int_c^d y g(y)\, dy$.

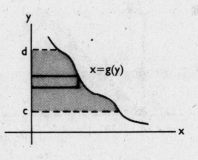

Q1. The region bounded by $y = x^3$, the x-axis, and the line $x = 2$ is revolved around the y-axis. Find the volume of the generated solid. _____

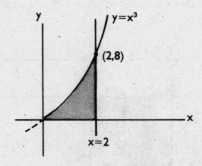

Answer: Using the shell method, $V = 2\pi \displaystyle\int_0^2 x \cdot x^3\, dx$

$$= 2\pi \int_0^2 x^4\, dx$$

$$= \frac{64\pi}{5} \text{ (cu. units)}.$$

Authors' Note: The "secret" to setting up the integration formula for the shell method is to realize that the volume of a shell is $2\pi r h t$, where $r = \dfrac{r_1 + r_2}{2}$, $h = \text{height}$, and $t = \text{thickness}$. Note the correspondence with the integration formula used in Q1.

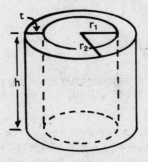

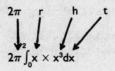

Problem Set 6.7

Use the shell method to calculate the volumes of the solids of revolution described.

1. The first quadrant region bounded by $y^2 = x$, the x-axis, and the line $x = 4$ revolved about the y-axis.

2. The region bounded by $y = e^x$, $y = 0$, $x = 0$, and $x = 2$ revolved about the y-axis.

3. The region bounded by $y = x^2$ and $y = x$ revolved about (a) the x-axis; (b) the y-axis.

4. The region bounded by $y = x^{3/2}$, the x-axis, and the line $x = 4$ revolved about the x-axis.

5. The region bounded by $y = \sin x$, the x-axis, and

the lines $x = 0$ and $x = \frac{\pi}{2}$ revolved about the y-axis.

6. The region bounded by $x = y^2 - 2y$ and $y = x$ revolved about the x-axis.

Solutions and Comments for Problem Set 6.7

1. $2\pi \int_0^4 x \cdot x^{1/2}\, dx = 2\pi \int_0^4 x^{3/2}\, dx$

$= \dfrac{128}{5}$ (cu. units).

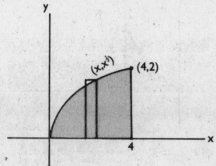

2. $2\pi \int_0^2 xe^x\, dx = 2\pi(e^2 + 1)$ (cu. units).

> Integration by parts yields
>
> $\int xe^x\, dx = xe^x - e^x + c.$

3. (a) $2\pi \int_0^1 y(y^{1/2} - y)\, dy = \dfrac{2\pi}{15}$ (cu. units);

   (b) $2\pi \int_0^1 x(x - x^2)\, dx = \dfrac{\pi}{6}$ (cu. units).

> Compare to problem #8 in the previous section. This will give a comparison of the ring method to the shell method.

4. $2\pi \int_0^8 y(4 - x)\, dy = 2\pi \int_0^8 y(4 - y^{2/3})\, dy$

$= 64\pi$ (cu. units).

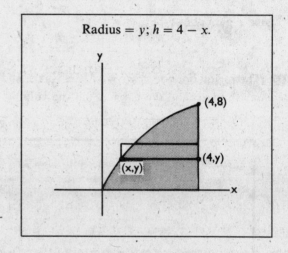

Radius $= y;\ h = 4 - x.$

5. $2\pi \int_0^{\pi/2} x \sin x\, dx = 2\pi$ (cu. units).

> $\int x \sin x\, dx = -x \cos x + \sin x + C$
>
> using integration by parts.

6. $2\pi \int_0^3 y[\,y - (y^2 - 2y)\,]\, dy$

$= 2\pi \int_0^3 (3y^2 - y^3)\, dy = \dfrac{27\pi}{2}$ (cu. units).

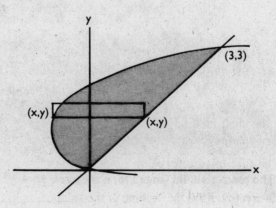

Using (X, y) and (x, y) in the figure, a typical radius of a shell would be y and a typical height $X - x$. The integral can be expressed as $2\pi \int_0^3 y(X - x)\, dx$. Note that it would be somewhat awkward to use the ring method in this problem.

*Section 6.7A: Volumes of Solids with Known Cross Sections

It is possible to find the volume of a solid (not necessarily a solid of revolution) by integration techniques if parallel cross sections obtained by slicing the solid with parallel planes perpendicular to an axis all have the same basic shape.

In the figure, a solid of the type described above lies between the planes $x = a$ and $x = b$. If each disk determined by two cross sections at $x = x_{i-1}$ and $x = x_i$ $(i = 1, 2, 3, \ldots, n)$ has height $\Delta x = \dfrac{b - a}{n}$ and if the area of the cross section at $x = x_i$ is $A(x_i)$, then the volume of the disk is approximately $A(x_i)\Delta x$.

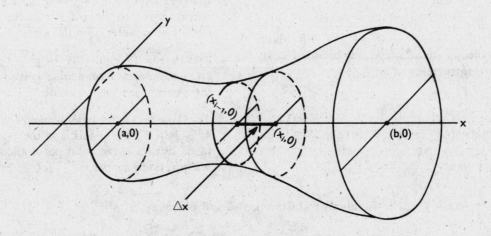

The volume of the entire solid is then

$$\lim_{n \to \infty} \sum_{i=1}^n A(x_i)\Delta x = \int_a^b A(x)\, dx.$$

Q1. The base of a solid (page 176) is the circle $x^2 + y^2 = 25$ and on each chord parallel to the y-axis a square is erected. Find the volume of the solid. _____

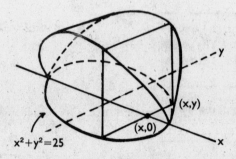

$x^2+y^2=25$

Answer: The area of the square containing $(x, 0)$ is $(2y)^2 = 4y^2 = 4(25 - x^2)$. Hence the desired volume is

$$\int_{-5}^{5} 4(25 - x^2)\, dx = 2 \int_{0}^{5} 4(25 - x^2)\, dx = 8\left(25x - \frac{x^3}{3}\right)\Big|_{0}^{5} = 666\tfrac{2}{3} \text{ (cu. units).}$$

Authors' Note: When evaluating $8\left(25x - \dfrac{x^3}{3}\right)\Big|_{0}^{5}$ one can save considerable computational time by noting

that $25x$ and $\dfrac{x^3}{3}$ both have three factors of 5 if $x = 5$. Hence, $8\left(25x - \dfrac{x^3}{3}\right)\Big|_{0}^{5} = 8 \times 5^3(1 - \tfrac{1}{3})$

$= 8 \times 5^3 \times \tfrac{2}{3}$.

Problem Set 6.7A

Find the volume of the solid described.

1. Base is the circle $x^2 + y^2 = 4$ and each plane perpendicular to the x-axis cuts the solid in an equilateral triangle.

2. Base is the circle $x^2 + y^2 = r^2$ and each plane perpendicular to the x-axis cuts the solid in an isosceles right triangle with one leg in the plane of the base.

3. Base is the circle $x^2 + y^2 = r^2$ and each plane perpendicular to the x-axis cuts the solid in a circle whose center is on the x-axis.

4. The base is the ellipse $4x^2 + y^2 = 1$ and every cross section is a semicircle perpendicular to the x-axis.

5. The base is a triangle formed by the line $4x + 3y = 12$, the x-axis, and the y-axis. Every plane section of the solid perpendicular to the x-axis is a semicircle.

Solutions and Comments for Problem Set 6.7A

1. $\sqrt{3} \displaystyle\int_{-2}^{2} (4 - x^2)\, dx = \dfrac{32\sqrt{3}}{3}$ (cu. units).

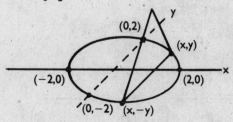

Area of triangle $= \dfrac{(2y)^2}{4}\sqrt{3}$

2. $2 \displaystyle\int_{-r}^{r} (r^2 - x^2)\, dx = \dfrac{8r^3}{3}$ (cu. units).

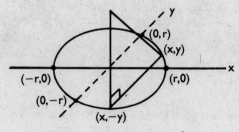

Area of triangle $= \tfrac{1}{2}(2y)^2$

3. $\int_{-r}^{r} \pi y^2 \, dx = 2 \int_{0}^{r} \pi(r^2 - x^2) \, dx = \frac{4}{3}\pi r^3$
(cu. units).

Area of semicircle $= \frac{1}{2}\pi y^2$

> We have derived the formula for the volume of a sphere of radius r.

5. $\frac{1}{2} \int_{0}^{3} \pi(2 - \frac{2}{3}x)^2 \, dx = 2\pi$ (cu. units).

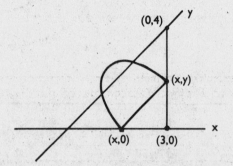

4. $\int_{-1/2}^{1/2} \frac{1}{2}\pi y^2 \, dx = \pi \int_{0}^{1/2} (1 - 4x^2) \, dx = \frac{\pi}{3}$
(cu. units).

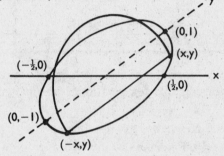

Area of semicircle $= \frac{1}{2}\pi(\frac{y}{2})^2$

*Section 6.8: Arc Length

If f is continuous on $[a, b]$, we can again divide the interval into n equal subintervals of length $\Delta x = \dfrac{b - a}{n}$ and form chords of arc AB as shown in the figure. If n is large, the length of arc AB is approximately the sum of the lengths of these chords. Note that

$$PQ = PR(\sec \angle RPQ) = \Delta x \sqrt{1 + \tan^2 \angle RPQ}.$$

The Mean Value Theorem (Section 3.9) assures us that there exists $x_i^* \in [x_{i-1}, x_i]$ such that

$$f'(x_i^*) = \frac{f(x_i) - f(x_{i-1})}{\Delta x} \quad (= \tan \angle RPQ).$$

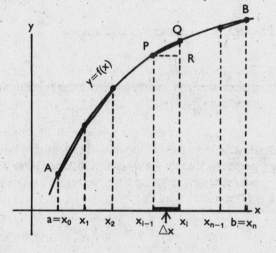

Hence $PQ = \Delta x \sqrt{1 + [f'(x_i^*)]^2}$ and the length of arc AB is approximately

$$\sum_{i=1}^{n} \sqrt{1 + [f'(x_i^*)]^2} \, \Delta x.$$

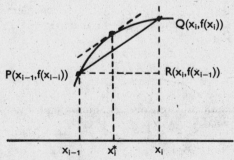

The actual length of arc AB is $\lim\limits_{n \to \infty} \sum\limits_{i=1}^{n} \sqrt{1 + [f'(x_i^*)]^2}\,\Delta x = \int_a^b \sqrt{1 + [f'(x)]^2}\,dx.$

Q1. Find the length of the arc of $y = \frac{2}{3}x^{3/2}$ between $x = 0$ and $x = 8$. _____

Answer: $\dfrac{dy}{dx} = x^{1/2}$ and $\left(\dfrac{dy}{dx}\right)^2 = x$. Hence the desired length is

$$\int_0^8 \sqrt{1 + x}\,dx = \frac{2}{3}(1 + x)^{3/2}\Big|_0^8 = \frac{2}{3}(27 - 1) = 17\frac{1}{3}.$$

When the curve is defined parametrically, $x = f(t)$, $y = g(t)$, then the length of the arc between $t = t_0$ and $t = t_1$ is

$$\int_{t_0}^{t_1} \sqrt{\left(\frac{dx}{dt}\right)^2 + \left(\frac{dy}{dt}\right)^2}\,dt.$$

Q2. The position of a particle at time t is given by $x = t$, $y = \frac{4}{3}t^{3/2}$. Find the distance it travels between $t = 0$ and $t = 2$. _____

Answer: $\dfrac{dx}{dt} = 1$ and $\dfrac{dy}{dt} = 2t^{1/2}$. The desired length is

$$\int_0^2 \sqrt{1 + 4t}\,dt = \frac{1}{6}(1 + 4t)^{3/2}\Big|_0^2 = \frac{1}{6}(27 - 1) = 4\frac{1}{3}.$$

Authors' Note: The reader should note that one might have to go to approximation methods for evaluating some integrals that could result from the arc length integration formulas. For example, the length of the arc of $y = \sin x$ between $x = 0$ and $x = \pi$ is given by

$$\int_0^\pi \sqrt{1 + \cos^2 x}\,dx.$$

There is no way to find an indefinite integral for $\sqrt{1 + \cos^2 x}$, but one could use the rectangular or trapezoidal approximation rules to find an approximate value for the arc length. The problems in the problem set will "work out" without the need for approximation formulas.

Problem Set 6.8

Calculate arc length over the designated interval in problems 1–6.

1. $y = x$ between $x = 0$ and $x = 4$

2. $y = x^{3/2}$ between $x = 0$ and $x = 4/3$

3. $y^2 = \frac{4}{9}(1 + x^2)^3$ between $x = 0$ and $x = 4$

4. $y = \frac{1}{2}(e^x + e^{-x})$ from $x = \ln 2$ to $x = \ln 4$

5. $f(x) = \ln \cos x$ on $\left[0, \frac{\pi}{4}\right]$

6. $x = e^{-t}\cos t$, $y = e^{-t}\sin t$ between $t = 0$ and $t = \pi$

7. The position of a particle at time t is given by $x = t^2$, $y = \frac{1}{3}t^3 - t$. Find the distance it travels between $t = 2$ and $t = 5$.

8. The parametric equations for a circle of radius r with center at the origin are $x = r\cos t$, $y = r\sin t$. Use the arc length formula and derive the formula for the circumference of this circle.

Solutions and Comments for Problem Set 6.8

1. $\int_0^4 \sqrt{1 + 1^2}\,dx = \int_0^4 \sqrt{2}\,dx = 4\sqrt{2}.$

> We are merely finding the length of the hypotenuse of an isosceles right triangle with legs of length 4. This problem illustrates the validity of the arc length formula.

2. $\int_0^{4/3} \sqrt{1 + \frac{9}{4}x}\,dx = \frac{8}{27}(1 + \frac{9}{4}x)^{3/2}\Big|_0^{4/3} = \frac{56}{27}.$

> Letting $u = 1 + \frac{9}{4}x$, the integral reduces to $\frac{4}{9}\int_1^4 u^{1/2}\,du$. (Note the changes in the upper and lower integration limits.)

3. $\frac{dy}{dx} = (1 + x^2)^{1/2}(2x)$. $\int_0^4 \sqrt{1 + 4x^2 + 4x^4}\,dx$
$= \int_0^4 (1 + 2x^2)\,dx = 46\frac{2}{3}.$

4. $\frac{dy}{dx} = \frac{e^x - e^{-x}}{2}$ and
$\sqrt{1 + \left(\frac{dy}{dx}\right)^2} = \sqrt{\frac{e^{2x} + 2 + e^{-2x}}{4}} = \frac{e^x + e^{-x}}{2}.$

$\int_{\ln 2}^{\ln 4} \frac{e^x + e^{-x}}{2}\,dx = \frac{e^x - e^{-x}}{2}\Big|_{\ln 2}^{\ln 4} = \frac{9}{8}.$

5. $f'(x) = \frac{1}{\cos x}(-\sin x) = -\tan x.$
$\int_0^{\pi/4} \sqrt{1 + \tan^2 x}\,dx = \int_0^{\pi/4} \sec x\,dx$
$= \ln|\sec x + \tan x|\Big|_0^{\pi/4}$
$= \ln(1 + \sqrt{2}).$

6. $\frac{dx}{dt} = -e^{-t}(\sin t + \cos t)$ and
$\frac{dy}{dt} = e^{-t}(\cos t - \sin t).$
$\sqrt{\left(\frac{dx}{dt}\right)^2 + \left(\frac{dy}{dt}\right)^2} = \sqrt{2}e^{-t}.$
$\sqrt{2}\int_0^\pi e^{-t}\,dt = \sqrt{2}(-e^{-t})\Big|_0^\pi = \sqrt{2}\left(1 - \frac{1}{e^\pi}\right).$

7. $\sqrt{\left(\frac{dx}{dt}\right)^2 + \left(\frac{dy}{dt}\right)^2} = \sqrt{(t^2 + 1)^2}.$
$\int_2^5 (t^2 + 1)\,dt = \left(\frac{t^3}{3} + t\right)\Big|_2^5 = 42.$

8. $\sqrt{r^2\sin^2 t + r^2\cos^2 t} = \sqrt{r^2} = r.$
$\int_0^{2\pi} r\,dt = rt\Big|_0^{2\pi} = 2\pi r.$

*Section 6.9: Area of a Surface of Revolution

If f is continuous on $[a, b]$, we can divide the interval into n subintervals of length $\Delta x = \frac{b - a}{n}$. If we rotate f about the x-axis, the resulting solid has a surface area which can be found by adding up the areas of n strips, one of which is displayed in the figure at the top of page 180.

Authors' Note: It may help to think of the strip as being
cut along $y = f(x)$ and rolled out flat
(see second figure).

As shown in Section 6.8, the length of arc ΔL is approxi-
mately $\Delta x \sqrt{1 + [f'(x_i^*)]^2}$, where x_i^* is a point in $[x_{i-1}, x_i]$
guaranteed by the Mean Value Theorem. The radius of
one base of the strip is $f(x_i)$ and the circumference of the
base is $2\pi f(x_i)$.

Hence the area of the strip is approximately

$$2\pi f(x_i)\sqrt{1 + [f'(x_i^*)]^2}\Delta x$$

and the actual surface area is

$$\lim_{n \to \infty} \sum_{i=1}^{n} 2\pi f(x_i)\sqrt{1 + [f'(x_i^*)]^2}\Delta x$$

$$- 2\pi \int_a^b f(x)\sqrt{1 + [f'(x)]^2}\, dx.$$

Q1. Find the area of the surface generated by rotating $y = x^3$, $0 \le x \le 1$ about the x-axis.

Answer: $\dfrac{dy}{dx} = 3x^2$. The desired area is

$$2\pi \int_0^1 x^3 \sqrt{1 + 9x^4}\, dx = \frac{\pi}{27}(1 + 9x^4)^{3/2}\Big|_0^1 = \frac{\pi}{27}(10^{3/2} - 1) \text{ (sq. units)}.$$

Authors' Note: In Q1, if $u = 1 + 9x^4$, then $du = 36x^3$. Also, $u = 10$ when $x = 1$ and $u = 1$ when $x = 0$.
The desired area can be obtained by calculating $2\pi \displaystyle\int_1^{10} \frac{1}{36}u^{1/2}\, du$.

If a portion of $x = g(y)$ between $y = c$ and $y = d$ is rotated about the y-axis, the surface area is

$$2\pi \int_c^d g(y)\sqrt{1 + [g'(y)]^2}\, dy.$$

When the curve is given parametrically, $x = f(t)$, $y = g(t)$, the surface area obtained by rotating the portion
between $t = t_0$ and $t = t_1$ about the x-axis is

$$2\pi \int_{t_0}^{t_1} y \sqrt{\left(\frac{dx}{dt}\right)^2 + \left(\frac{dy}{dt}\right)^2}\, dt.$$

Q2. The curve defined by $x = t^2$, $y = \frac{1}{3}t^6$ $(0 \le t \le 1)$ is rotated about the x-axis. Find the surface area of the generated solid. _____

$$\text{Answer: } 2\pi \int_0^1 \frac{t^6}{3}\sqrt{(2t)^2 + (2t^5)^2}\, dt = \frac{4\pi}{3}\int_0^1 t^7\sqrt{1 + t^8}\, dt$$

$$= \frac{\pi}{9}(1 + t^8)^{3/2}\Big|_0^1 = \frac{\pi}{9}(2^{3/2} - 1) \text{ (sq. units).}$$

Authors' Note: We would obtain the same result by rotating $y = \dfrac{x^3}{3}$, $0 \le x \le 1$ about the x-axis.

Problem Set 6.9

1. $y = 2x$ from $x = 0$ to $x = 4$ is rotated about the x-axis. Find the surface area of the resulting solid.

2. $y = 2x^{1/2}$, $0 \le x \le 3$ is rotated about the x-axis. Find the surface area.

3. $y = \frac{1}{2}(e^x + e^{-x})$, $0 \le x \le 1$ is rotated about the x-axis. Find the surface area.

4. The curve $y = x^{1/3}$, $0 \le x \le 8$ is rotated about the y-axis. Find the surface area.

5. The parametric equations $x = r\cos t$ and $y = r\sin t$ represent a circle with center at the origin and radius r. Find the surface area of the sphere generated when the circle is rotated about the x-axis.

6. The curve defined by $x = t - \sin t$, $y = 1 - \cos t$ between $t = 0$ and $t = \pi$ is rotated about the x-axis. Find the surface area of the generated solid.

Solutions and Comments for Problem Set 6.9

1. $2\pi \displaystyle\int_0^4 2x\sqrt{1 + 2^2}\, dx = 32\pi\sqrt{5}$ (sq. units).

> Solid of revolution is a right circular cone with base radius 8 and altitude 4. The slant height is $4\sqrt{5}$. Surface area of cone (excluding base) is $\pi r l$. Note that this formula produces the same answer as the calculus process.

2. $\dfrac{dy}{dx} = x^{-1/2}$. $2\pi \displaystyle\int_0^3 2x^{1/2}\sqrt{1 + x^{-1}}\, dx$

$= 4\pi \displaystyle\int_0^3 \sqrt{x + 1}\, dx = \dfrac{8\pi}{3}(x + 1)^{3/2}\Big|_0^3$

$= \dfrac{56\pi}{3}$ (sq. units).

> Don't confuse $2x^{1/2}$ and $(2x)^{1/2}$, a common error.

3. $\dfrac{dy}{dx} = \dfrac{1}{2}(e^x - e^{-x})$ and $\sqrt{1 + \left(\dfrac{dy}{dx}\right)^2} =$

$\dfrac{1}{2}(e^x + e^{-x})$. $2\pi \displaystyle\int_0^1 \dfrac{1}{2}(e^x + e^{-x}) \times$

$\dfrac{1}{2}(e^x + e^{-x})\, dx = \dfrac{\pi}{2}\displaystyle\int_0^1 (e^{2x} + 2 + e^{-2x})\, dx =$

$\dfrac{\pi}{2}\left(\dfrac{e^{2x}}{2} + 2x - \dfrac{e^{-2x}}{2}\right)\Big|_0^1 =$

$\dfrac{\pi(e^2 - e^{-2} + 4)}{4}$ (sq. units).

4. $x = y^3$. If $x = 0$, then $y = 0$; if $x = 8$, then $y = 2$.

In this instance, we need $\dfrac{dx}{dy}$ $\left(\text{not } \dfrac{dy}{dx}\right)$; $\dfrac{dx}{dy} = 3y^2$

and $\sqrt{1 + \left(\dfrac{dx}{dy}\right)^2} = \sqrt{1 + 9y^4}$.

$2\pi \displaystyle\int_0^2 y^3 \sqrt{1 + 9y^4}\, dy = \dfrac{\pi}{27}(1 + 9y^4)^{3/2} \Big|_0^2 =$

$\dfrac{\pi}{27}(145^{3/2} - 1)$ (sq. units).

5. $2\pi \displaystyle\int_0^\pi r \sin t \sqrt{(-r \sin t)^2 + (r \cos t)^2}\, dt =$

$2\pi \displaystyle\int_0^\pi r^2 \sin t\, dt = 2\pi r^2(-\cos t)\Big|_0^\pi =$

$4\pi r^2$ (sq. units).

> We have derived the famous formula for the surface area of a sphere of radius r.

6. $2\pi \displaystyle\int_0^\pi (1 - \cos t)\sqrt{2 - 2\cos t}\, dt =$

$2\pi \displaystyle\int_0^\pi 2\dfrac{(1 - \cos t)}{2} \times 2\sqrt{\dfrac{1 - \cos t}{2}}\, dt =$

$8\pi \displaystyle\int_0^\pi \sin^3 \dfrac{t}{2}\, dt = \dfrac{32\pi}{3}$ (sq. units).

> This one is a bit tricky. Remember that $\sin\dfrac{\theta}{2} = \sqrt{\dfrac{1 - \cos\theta}{2}}$ and recall that there are reduction formulas for $\displaystyle\int \sin^n \theta\, d\theta$ and $\displaystyle\int \cos^n \theta\, d\theta$.

*Section 6.10: Improper Integrals

In attempting to evaluate $\displaystyle\int_a^b f(x)\, dx$ we have required that f be continuous on $[a, b]$. However, there are two situations that often produce interesting results when f is not continuous throughout the entire closed interval. There are when

(1) a or b is infinite, or

(2) f is discontinuous at a point in $[a, b]$.

If (1) or (2) is true, then the integral is an *improper integral*.

In situation (1),

$$\int_a^\infty f(x)\, dx = \lim_{b \to \infty} \int_a^b f(x)\, dx$$ if this limit exists. If the limit does exist, then the integral is said to be *convergent*; otherwise, it is *divergent*.

Similarly,

$$\int_{-\infty}^b f(x)\, dx = \lim_{a \to -\infty} \int_a^b f(x)\, dx$$ if this limit exists, and

$$\int_{-\infty}^\infty f(x)\, dx = \int_{-\infty}^0 f(x)\, dx + \int_0^\infty f(x)\, dx$$ if and only if both of the integrals (limits) on the right-hand side of the equation exist.

Q1. Evaluate $\displaystyle\int_1^\infty x^{-3/2}\, dx$. _____

Answer: $\lim\limits_{b\to\infty}\displaystyle\int_1^b x^{-3/2}\,dx = \lim\limits_{b\to\infty}(-2x^{-1/2})\Big|_1^b = \lim\limits_{b\to\infty}\left(-\dfrac{2}{b^{1/2}}+2\right) = 0+2 = 2.$
The improper integral is convergent.

Authors' Note: Geometrically, the area bounded by $x = 1$, $y = x^{-3/2}$, and the portion of the x-axis where $x \geq 1$, while infinite in extent, does not exceed 2 sq. units.

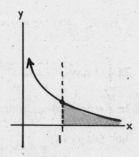

Q2. Evaluate $\displaystyle\int_0^\infty e^x\,dx.$ _____

Answer: $\lim\limits_{b\to\infty}\displaystyle\int_0^b e^x\,dx = \lim\limits_{b\to\infty} e^x\Big|_0^b = \lim\limits_{b\to\infty}(e^b - 1) = \infty.$ The improper integral is divergent.

Now consider situation (2):

If f is continuous on $[a, b)$ with $f(b) = \infty$ or $-\infty$, then

$$\int_a^b f(x)\,dx = \lim_{x\to b^-}\int_a^x f(x)\,dx \text{ if this limit exists. (Remember that } x\to b^- \text{ means "as } x \text{ approaches } b \text{ from the left.")}$$

If f is continuous on $(a, b]$ with $f(a) = \infty$ or $-\infty$, then

$$\int_a^b f(x)\,dx = \lim_{x\to a^+}\int_x^b f(x)\,dx \text{ if this limit exists.}$$

If $c \in (a, b)$ and f is continuous on $[a, b]$ except at $x = c$, then

$$\int_a^b f(x)\,dx = \lim_{x\to c^-}\int_a^x f(x)\,dx + \lim_{x\to c^+}\int_x^b f(x)\,dx.$$

Q3. Evaluate $\displaystyle\int_2^3 \dfrac{dx}{(3-x)^{3/4}}.$ _____

Answer: $\lim\limits_{b\to 3}\displaystyle\int_2^b \dfrac{dx}{(3-x)^{3/4}} = \lim\limits_{b\to 3} -4(3-x)^{1/4}\Big|_2^b = 0+4 = 4.$

Authors' Note: As a contrast to Q3, note that $\int_2^3 \dfrac{dx}{(3-x)^{4/3}} = \infty$.

 Common error: $\int_{-2}^2 \dfrac{dx}{x^2} = -\dfrac{1}{x}\Big|_{-2}^2 = -\dfrac{1}{2} - \left(-\dfrac{1}{-2}\right) = -1.$

 Authors' Note: $f(x) = \dfrac{1}{x^2}$ is discontinuous at $x = 0$. This does not mean that the improper integral diverges, but, in this case, if we write $\int_{-2}^2 \dfrac{dx}{x^2} = \lim_{b\to 0^-} \int_{-2}^b \dfrac{dx}{x^2} + \lim_{a\to 0^+} \int_a^2 \dfrac{dx}{x^2}$, neither of the limits on the right exist. Hence $\int_{-2}^2 \dfrac{dx}{x^2}$ is divergent.

 Common error: $\int_{-\infty}^\infty x\,dx = \lim_{b\to\infty} \int_{-b}^b x\,dx = \lim_{b\to\infty} \dfrac{x^2}{2}\Big|_{-b}^b = \lim_{b\to\infty}\left[\dfrac{b^2}{2} - \dfrac{b^2}{2}\right] = \lim_{b\to\infty} 0 = 0.$

 Authors' Note: This is a "subtle" error. Note that $\int_{-\infty}^\infty x\,dx$ is *not* defined to be $\lim_{b\to\infty} \int_{-b}^b x\,dx$, but rather as $\int_{-\infty}^0 x\,dx + \int_0^\infty x\,dx$ and neither of the two limits in the sum exists. The expression $\infty - \infty$, which (loosely) one could obtain as the indicated sum, is not defined (and certainly is not zero). The improper intergal $\int_{-\infty}^\infty x\,dx$ diverges.

Problem Set 6.10

1. Find the area bounded by $x \geq 1$, the x-axis, and the curve
   (a) $y = \dfrac{1}{x^2}$
   (b) $y = \dfrac{1}{x}$

2. Evaluate $\int_{-\infty}^0 e^x\,dx$.

3. Evaluate $\int_1^9 \dfrac{dx}{(x-1)^{2/3}}$.

4. Evaluate $\int_1^\infty \dfrac{dx}{x^{1/3}}$.

5. Evaluate $\int_0^\infty \cos x\,dx$.

6. Evaluate $\int_0^3 \dfrac{dx}{(x-1)^{2/3}}$.

7. Evaluate $\int_{-\infty}^\infty \dfrac{1}{1+x^2}\,dx$.

8. Calculate the area in the first quadrant under the curve $y = x^{-1/2}$ between $x = 0$ and $x = 1$.

9. Calculate $\int_{-1}^8 \dfrac{1}{\sqrt[3]{x}}\,dx$.

Solutions and Comments for Problem Set 6.10

1. (a) $\lim_{b\to\infty}\left(-\frac{1}{x}\right)\Big|_1^b = 1$ (sq. unit);

   (b) $\lim_{b\to\infty}\ln x\Big|_1^b = \infty$.

> This problem illustrates the distinction between convergent and divergent improper integrals.

2. $\lim_{a\to-\infty} e^x\Big|_a^0 = 1$.

> Compare to Q2 in the reading section.

3. $\lim_{a\to1^+}\int_a^9\frac{dx}{(x-1)^{2/3}} = \lim_{a\to1^+} 3(x-1)^{1/3}\Big|_a^9$
   $= 6 - 0 = 6$.

4. $\lim_{b\to\infty}\frac{3}{2}x^{2/3}\Big|_1^b = \infty$. Improper integral diverges.

5. $\lim_{b\to\infty}\sin x\Big|_0^b = \lim_{b\to\infty}\sin b$. Limit does not exist.

> Note that $\sin b$ takes on values between -1 and 1 as b gets larger and larger. That is, the value of the expression does not approach a given number (and does not become infinite).

6. $\lim_{b\to1^-}\int_0^b\frac{dx}{(x-1)^{2/3}} + \lim_{a\to1^+}\int_a^3\frac{dx}{(x-1)^{2/3}} =$
   $3 + 3\sqrt[3]{2}$.

> Compare to problem #3.

7. $\lim_{a\to-\infty}(\tan^{-1} x)\Big|_a^0 + \lim_{b\to\infty}(\tan^{-1} x)\Big|_0^b = \frac{\pi}{2} + \frac{\pi}{2} = \pi$.

> Remember the restrictions on the inverse trig functions. In this case, $-\frac{\pi}{2} < \tan^{-1} x < \frac{\pi}{2}$.

8. $\lim_{a\to0^+} 2\sqrt{x}\Big|_a^1 = 2 - 0 = 2$ (sq. units).

> Region is infinite, but has a finite area.

9. $\lim_{b\to0^-}\frac{3}{2}x^{2/3}\Big|_{-1}^b + \lim_{a\to0^+}\frac{3}{2}x^{2/3}\Big|_a^8 = -\frac{3}{2} + 6 = 4\frac{1}{2}$.

> Graph has vertical asymptote at $x = 0$.

Section 6.11: Multiple Choice Questions for Chapter 6

1. $\int_{\ln 2}^{\ln 9} 4e^x\,dx =$
   (A) e^7
   (B) 28
   (C) 72
   (D) 32
   (E) $4\ln(9/2)$

2. $\int_{\pi/3}^{\pi/4}\tan x\sin x\cot x\csc x\,dx =$
   (A) $\pi/6$
   (B) 1
   (C) $-\pi/6$
   (D) $\pi/12$
   (E) $-\pi/12$

3.

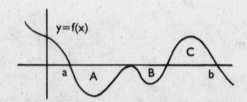

If *A*, *B*, and *C* represent the areas of the regions shown in the figure, then $\int_a^b f(x)\,dx =$

(A) $A + B - C$
(B) $A - B + C$
(C) $C - A - B$
(D) $-A + C + B$
(E) $C - B + A$

4. $\lim_{n \to \infty} \dfrac{1}{n} \sum_{k=n+1}^{2n} \dfrac{k}{n} =$
(A) 1/2
(B) 1
(C) 5/4
(D) 3/2
(E) 2

5. *How many* of the following statements are true?

(i) $\int_a^b f(kx)\,dx = k \int_a^b f(x)\,dx$

(ii) $\int_p^w g(x)\,dx = -\int_w^p g(x)\,dx$

(iii) $\int_{-a}^a f(x)\,dx = 2 \int_0^a f(x)\,dx$

(iv) $\int_c^c u(x)\,dx = 0$

(A) 0
(B) 1
(C) 2
(D) 3
(E) 4

6. $\int_{-2}^2 |x|\,dx =$
(A) 4
(B) 1
(C) 0
(D) 1/2
(E) 8

7. A function f is continuous on $[13, 16]$ and $3 \le f(x) \le 7$ for all $x \in [13, 16]$. If $p \le \int_{13}^{16} f(x)\,dx \le q$, then the minimum value for p and the maximum value for q are, respectively,
(A) 52, 64
(B) 9, 21
(C) 3, 7
(D) 39, 48
(E) 0, 64

8. $\dfrac{d}{dx} \int_7^x w \sin w \, dw =$
(A) $x \sin x$
(B) $\cos x$
(C) $\sin x - x \cos x$
(D) $x^2 \sin^2 x$
(E) $\dfrac{x^2}{4} \sin^2 x$

9. $\dfrac{d}{dx} \int_0^{x^3} \cos^4 t \, dt =$
(A) $\cos^4 x$
(B) $\cos^4 (x^3)$
(C) $\sin^4 (x^3)$
(D) $\dfrac{\cos^5 (x^3)}{5}$
(E) $3x^2 \cos^4 (x^3)$

10. $\int_0^4 \dfrac{2x}{x^2 + 9}\,dx =$
(A) 25
(B) 16
(C) $\ln(25/9)$
(D) $\ln 4$
(E) $\ln(5/3)$

11. Find the area of the region bounded by the graph of $y = x^2$, $x = 5$, $x = 2$, and $y = 0$.
(A) 27 sq. units
(B) 39 sq. units
(C) $42\frac{1}{3}$ sq. units
(D) 45 sq. units
(E) $52\frac{1}{3}$ sq. units

12.

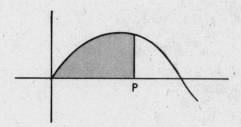

The graph shows a portion of $y = \sin x$ from $x = 0$ to $x = p$. If the area of the shaded region is 0.8 sq. units, what is the value of $\cos p$?
(A) 1/2
(B) $\pi/6$
(C) $\pi/3$
(D) 2/5
(E) 1/5

13. What is the volume of the solid obtained by rotating the region bounded by $y = 1 - x^2$ and $y = 0$ about the x-axis?
(A) 2π cu. units
(B) $\dfrac{16\pi}{15}$ cu. units
(C) $\dfrac{5\pi}{3}$ cu. units
(D) $\dfrac{12\pi}{7}$ cu. units
(E) π cu. units

14. Which definite integral represents the volume of a sphere with radius 5?
(A) $\pi \displaystyle\int_{-5}^{5} (x^2 + 25)\,dx$
(B) $\pi \displaystyle\int_{0}^{5} (25 - x^2)\,dx$
(C) $\pi \displaystyle\int_{-5}^{5} (5 - x^2)\,dx$
(D) $2\pi \displaystyle\int_{0}^{5} (25 - x^2)\,dx$
(E) $\pi \displaystyle\int_{-5}^{5} (x^2 - 25)\,dx$

15. The region bounded by $y = 3x$, $x = 0$, $x = 3$, and $y = 0$ is rotated about the y-axis. Using the shell method, the volume of the solid generated is represented by
(A) $\pi \displaystyle\int_{0}^{9} xy\,dx$

(B) $2\pi \displaystyle\int_{0}^{3} xy\,dy$
(C) $2\pi \displaystyle\int_{0}^{3} xy\,dx$
(D) $2\pi \displaystyle\int_{0}^{3} xy^2\,dx$
(E) $\pi \displaystyle\int_{0}^{3} xy\,dy$

16. $\displaystyle\int_{4}^{25} \dfrac{e^{\sqrt{x}}}{\sqrt{x}}\,dx =$
(A) $2e^2(e^3 - 1)$
(B) $e^5 - e^2$
(C) $\frac{1}{2}(e^5 - e^2)$
(D) $e^4(e^{21} - 1)$
(E) $\frac{1}{2}e^5 - e^2$

17. $\displaystyle\int_{\sqrt{3}/3}^{1} \dfrac{dx}{1 + x^2} =$
(A) $\pi/6$
(B) $\pi/12$
(C) $\pi/2$
(D) $-\pi/2$
(E) $\pi/3$

*18. Find the area of the region bounded by the polar coordinate graphs of $r = \sqrt{\cos\theta}$, $\theta = 0$ and $\theta = \pi/2$.
(A) $\pi/8$
(B) $\pi/2$
(C) 1/2
(D) 4/3
(E) $\pi/3$

*19. Find the length of $y = x^{3/2}$ from $x = 0$ to $x = 5$.
(A) 335/27
(B) 221/27
(C) 87/9
(D) $6\frac{1}{2}$
(E) $8\frac{2}{3}$

*20. The region bounded by $y = 2\sqrt{x}$, $x = 3$, $x = 15$, and $y = 0$ is rotated about the x-axis. What is the area of the surface generated?
(A) 168π
(B) $332\pi/3$
(C) $614\pi/9$
(D) $448\pi/3$
(E) $319\pi/2$

*21. $\int_{6}^{10} \dfrac{dx}{\sqrt{x-6}} =$

(A) 8
(B) 6
(C) 5
(D) 4
(E) integral diverges

*22. If the trapezoidal rule is applied to $\int_{2}^{3} x^3\, dx$ with $\Delta x = \frac{1}{2}$, the approximate value for the integral is

(A) $22\frac{1}{3}$
(B) $20\frac{1}{4}$
(C) $19\frac{3}{4}$
(D) $17\frac{1}{8}$
(E) $16\frac{9}{16}$

*23. $\int_{0}^{\infty} \dfrac{dx}{\sqrt{x+1}} =$

(A) 1

(B) $4\frac{2}{9}$
(C) 87/7
(D) $14\frac{2}{3}$
(E) integral diverges

*24. $\int_{1}^{\infty} \dfrac{dx}{1+x^2} =$

(A) 1
(B) $\pi/4$
(C) $\pi/2$
(D) π
(E) integral diverges

*25. The base of a solid is the circle $x^2 + y^2 = 9$ and each plane perpendicular to the x-axis cuts the solid in an equilateral triangle. Find the volume of the solid.

(A) $36\sqrt{3}$
(B) $24\sqrt{3}$
(C) $18\sqrt{3}$
(D) $\sqrt{18}$
(E) $12\sqrt{3}$

Solutions and Comments for Multiple Choice Questions

1. **(B)** $4e^x \Big|_{\ln 2}^{\ln 9} = 4(9-2) = 28$.

> $e^{\ln t} = t.$

2. **(E)** $(\tan x)(\cot x)(\sin x)(\csc x) = 1 \times 1 = 1$ on $[\pi/4, \pi/3]$. $\int_{\pi/3}^{\pi/4} dx = x \Big|_{\pi/3}^{\pi/4} = \dfrac{\pi}{4} - \dfrac{\pi}{3} = -\dfrac{\pi}{12}$.

3. **(C)** Section 6.5.

4. **(D)** This is just $\int_{1}^{2} x\, dx$.

5. **(C)** (ii) and (iv) are true.

> If you are careless, you can easily confuse (i) with property (2) in Section 6.3.

6. **(A)** This is $2\int_{0}^{2} x\, dx$. $f(x) = |x|$ is an even function.

7. **(B)** $3(16-13) \le \int_{13}^{16} f(x)\, dx \le 7(16-13)$.

> A direct application of (6) in Section 6.3.

8. **(A)** $\dfrac{d}{dx} \int_{a}^{x} f(x)\, dx = f(x)$.

9. **(E)** This is an application of the second highlighted formula in Section 6.4 with $h(x) = 0$.

10. **(C)** $\ln(x^2 + 9) \Big|_{0}^{4} = \ln 25 - \ln 9$.

> If $u = x^2 + 9$, then $du = 2x$ and the integral becomes $\int_{9}^{25} \dfrac{du}{u}$.

11. **(B)** $\int_{2}^{5} x^2\, dx$.

12. **(E)** $\int_0^p \sin x \, dx = -\cos x \Big|_0^p = -\cos p + 1 =$
$0.8 \Rightarrow \cos p = 0.2$.

13. **(B)** Using the disk method,
$$\int_{-1}^1 \pi(1 - x^2)^2 \, dx = \int_{-1}^1 \pi(1 - 2x^2 + x^4) \, dx$$
$$= \pi\left(x - \frac{2}{3}x^3 + \frac{x^5}{5}\right)\Big|_{-1}^1 = \frac{16\pi}{15}.$$

14. **(D)**

> **(D)** represents the volume of the sphere formed when the semicircle $y = \sqrt{25 - x^2}$ is revolved around the x-axis.

15. **(C)** Evaluate $2\pi \int_0^3 x(3x) \, dx$.

16. **(A)** If $u = \sqrt{x}$, then $x = u^2$ and $dx = 2u \, du$. $u = 5$ when $x = 25$; $u = 2$ when $x = 4$. Hence, integral becomes $\int_2^5 \frac{e^u}{u} 2u \, du = 2 \int_2^5 e^u \, du =$
$2e^u \Big|_2^5 = 2e^5 - 2e^2$.

17. **(B)** $\int \frac{dx}{1 + x^2} = \tan^{-1} x + C$.
$\tan^{-1} x \Big|_{\sqrt{3}/3}^1 = \frac{\pi}{4} - \frac{\pi}{6} = \frac{\pi}{12}$.

18. **(C)** $\frac{1}{2} \int_0^{\pi/2} \cos \theta \, d\theta = \frac{1}{2} \sin \theta \Big|_0^{\pi/2} = \frac{1}{2}$.
Section 6.5A.

19. **(A)** $f'(x) = \frac{3}{2} x^{1/2}$ and $[f'(x)]^2 = \frac{9}{4} x$.
$\int_0^5 \sqrt{1 + \frac{9}{4}x} \, dx = \frac{335}{27}$. (Let $u = 1 + \frac{9}{4}x$.)

20. **(D)** If $y = 2x^{1/2}$, then $\frac{dy}{dx} = x^{-1/2}$ and
$\left(\frac{dy}{dx}\right)^2 = x^{-1}$. $2\pi \int_3^{15} 2\sqrt{x}\sqrt{1 + x^{-1}} \, dx =$
$4\pi \int_3^{15} (x + 1)^{1/2} \, dx = \frac{8\pi}{3}(x + 1)^{3/2} \Big|_3^{15} =$
$\frac{8\pi}{3}(16^{3/2} - 4^{3/2}) = \frac{448\pi}{3}$.

21. **(D)** $\lim_{b \to 6^+} \int_b^{10} \frac{dx}{\sqrt{x - 6}} = \lim_{b \to 6^+} 2\sqrt{x - 6} \Big|_b^{10}$
$= 2 \times 2 - 2 \times 0 = 4$.

> Improper integral converges.

22. **(E)** $\frac{1}{2}\left[\frac{1}{2}f(2) + f\left(\frac{5}{2}\right) + \frac{1}{2}f(3)\right] = \frac{265}{16}$.
Section 6.1A.

23. **(E)** $\lim_{b \to \infty} \int_0^b \frac{dx}{\sqrt{x + 1}} = \lim_{b \to \infty} 2\sqrt{x + 1} \Big|_0^b$
$= \lim_{b \to \infty} \left[2\sqrt{b + 1} - 2\right]$
$= \infty$.

> Improper integral diverges.

24. **(B)** $\lim_{b \to \infty} \int_1^b \frac{dx}{1 + x^2} = \lim_{b \to \infty} \tan^{-1} x \Big|_1^b$
$= \frac{\pi}{2} - \frac{\pi}{4} = \frac{\pi}{4}$.

> As $b \to \infty$, $\tan^{-1} b \to \frac{\pi}{2}$.

25. **(A)** $\int_{-3}^3 \sqrt{3}(9 - x^2) \, dx = 36\sqrt{3}$.

> Since $x^2 + y^2 = 9$, we have $y = \pm\sqrt{9 - x^2}$. Hence a side of an equilateral triangle is $2\sqrt{9 - x^2}$ and the altitude is $(\sqrt{9 - x^2})\sqrt{3}$. The area is $\frac{1}{2}(2\sqrt{9 - x^2})(\sqrt{9 - x^2})\sqrt{3} = \sqrt{3}(9 - x^2)$. See Section 6.7A.

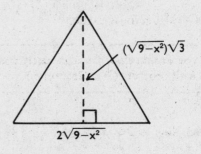

$(\sqrt{9-x^2})\sqrt{3}$

$2\sqrt{9-x^2}$

Chapter 7
SEQUENCES AND SERIES

---

*Section 7.1: Convergence and Divergence

Definition: A *sequence* is a function whose domain is the set of positive integers.

Usually only the range of the function is written and the sequence is expressed as a set of elements called *terms*; for example

$$\{a_1, a_2, a_3, \ldots, a_n\}.$$

The term corresponding to the *n*-th integer is a_n. This term is frequently used to define the sequence in one of two possible ways:

(1) As a *general term*, in which a_n is expressed as a function of *n*, or
(2) in *recursive form*, where a_n is expressed in terms of previous terms.

Example: Consider the sequence $\{1, 4, 7, 10, \ldots\}$.
The general term is $a_n = 3n - 2$, where $n = 1, 2, 3, \ldots$. A recursive form of this sequence is not unique, but one such form is

$$a_1 = 1$$
$$a_n = a_{n-1} + 3.$$

Authors' Note: In recursive form, the first term of the sequence must be identified. That is, there must be a "starting point" for the recursive pattern.

Finally, a sequence can be *finite* or *infinite*, but we will be concerned primarily with infinite sequences in this chapter.

If a sequence $\{a_n\}$ has a limit L $\left(\lim\limits_{n \to \infty} a_n = L \right)$, then the sequence is said to *converge* to L. If there is no limit, the sequence *diverges*.

Q1. Determine whether the sequence converges. If it does, find the limit.

(a) $a_n = \dfrac{n+1}{n-1}$ _____

(b) $a_n = (-1)^n \left[\dfrac{n+1}{n^2+1} \right]$ _____

(c) $a_n = \cos n$ _____

(d) $a_n = \dfrac{n^2+1}{n+1}$ _____

> Answers: (a) $\lim\limits_{n \to \infty} \dfrac{n+1}{n-1} = 1.$ (b) $\lim\limits_{n \to \infty} (-1)^n \left[\dfrac{n+1}{n^2+1} \right] = 0.$
>
> (c) no limit. (d) no limit.

Authors' Note: The reader should note that the simple techniques discussed in Chapter 2 can be utilized in many limit problems. And, of course, L'Hopital's Rule (Section 3.11) can be used in (a), (b), and (d).

A *series* can be represented by the sum of terms of a sequence.

An *infinite series* is often defined in terms of a sequence of partial sums in the following way. Consider the infinite sequence $\{a_1, a_2, a_3, \ldots a_n, \ldots\}$. Form the sequence of partial sums $s_1, s_2, s_3, \ldots s_n, \ldots$ such that

$s_1 = a_1$
$s_2 = a_1 + a_2$
$s_3 = a_1 + a_2 + a_3$
$s_n = a_1 + a_2 + a_3 + \cdots + a_n$

If the sequence of partial sums has a limit L, then the value of the infinite series $\sum\limits_{i=1}^{\infty} a_i$ is L and we say that it converges.

Authors' Note: Remember that a sequence is a *set* formed by a rule and has a one-to-one correspondence with the positive integers. A *series* is a *sum*. In effect, finding the limit of a series enables us to find the sum of an infinite number of terms.

Important Facts:
(1) A sequence or series is *increasing* if $a_{n+1} \geq a_n$ for all n.
(2) A sequence or series is *decreasing* if $a_{n+1} \leq a_n$ for all n.
(3) If the series converges, then a_n approaches 0 as n increases (goes to ∞).

Authors' Note: It is very important to realize that this is a *necessary* but not *sufficient* condition for convergence. The converse of this idea is *not* true, since it is possible for the series to diverge even when a_n goes to 0. The following example of the harmonic series demonstrates this clearly.

Example: The *harmonic series*

$$\sum_{k=1}^{\infty} \frac{1}{k} = 1 + \frac{1}{2} + \frac{1}{3} + \frac{1}{4} + \frac{1}{5} + \cdots$$

is often useful in comparison tests. Despite the fact that $\lim\limits_{n\to\infty} \dfrac{1}{n} = 0$, the series diverges as shown by the following proof: First, consider the harmonic series, grouped as follows:

$$1 + \frac{1}{2} + \left(\frac{1}{3} + \frac{1}{4}\right) + \left(\frac{1}{5} + \frac{1}{6} + \frac{1}{7} + \frac{1}{8}\right) + \left(\frac{1}{9} + \frac{1}{10} + \frac{1}{11} + \frac{1}{12} + \frac{1}{13} + \frac{1}{14} + \frac{1}{15} + \frac{1}{16}\right) + \cdots$$

It is clearly greater, term by term, then the following series:

$$1 + \frac{1}{2} + \left(\frac{1}{4} + \frac{1}{4}\right) + \left(\frac{1}{8} + \frac{1}{8} + \frac{1}{8} + \frac{1}{8}\right) + \left(\frac{1}{16} + \frac{1}{16} + \frac{1}{16} + \frac{1}{16} + \frac{1}{16} + \frac{1}{16} + \frac{1}{16} + \frac{1}{16}\right) + \cdots$$

Since the second series is an infinite sum of terms, each equal to $\frac{1}{2}$, it obviously diverges. Since, term by term, the harmonic series is greater, it must also diverge.

(4) A series of the form $\sum\limits_{i=1}^{\infty} ar^{i-1} = a + ar + ar^2 + ar^3 + \cdots$ is called a *geometric series*. The sum of the first n terms is $S_n = \dfrac{a(1 - r^n)}{1 - r}, r \neq 1$, and, if $|r| < 1$, the infinite series will converge to $\dfrac{a}{1 - r}$. If $|r| \geq 1$, the series diverges provided $a \neq 0$. If $r = 1$, the series is a sum of constants.

(5) Multiplying a series by a constant has no effect on the convergence or divergence.

(6) The convergence or divergence of a series is not affected by the addition or subtraction of any finite number of terms.

Q2. Express the repeating decimal $0.121212\ldots$ as a rational number by finding the sum of the infinite series it represents. _____

Answers: $0.\overline{12} = \dfrac{12}{100} + \dfrac{12}{10^4} + \dfrac{12}{10^6} + \cdots$.

$$a = \frac{12}{100}, r = \frac{1}{100}.$$

Hence $S = \dfrac{12/100}{1 - 1/100} = \dfrac{12}{99} = \dfrac{4}{33}$.

Problem Set 7.1

In problems 1–4, determine whether the sequence converges or diverges. If it converges, state the limit.

1. $a_n = \dfrac{n + 2n^2}{1 - n^2}$

2. $a_n = \dfrac{4n^2 - 1}{n^2 + 3}$

3. $a_n = \dfrac{3n - 2}{n^2 - 5}$

4. $a_n = \dfrac{1}{2^n}$

In problems 5–7, determine whether the indicated series converges.

5. $\sum\limits_{n=1}^{\infty} \dfrac{1}{2n}$

6. $1 + 2 + 3 + \sum\limits_{n=1}^{\infty} \dfrac{1}{2^n}$

7. $\sum\limits_{k=1}^{\infty} \dfrac{1}{k(k + 1)} = \sum\limits_{k=1}^{\infty} \left(\dfrac{1}{k} - \dfrac{1}{k + 1}\right)$

Solutions and Comments for Problem Set 7.1

1. $\lim_{n \to \infty} a_n = -2$.

> As shown earlier in the limits chapter, dividing by the highest power of n (in this case, dividing by n^2) yields the ratio -2.

2. $\lim_{n \to \infty} a_n = \lim_{n \to \infty} \dfrac{4 - \dfrac{1}{n^2}}{1 + \dfrac{3}{n^2}} = 4$.

3. $\lim_{n \to \infty} a_n = 0$.

> The degree of the polynomial in the numerator is less than the degree of the polynomial in the denominator.

4. $\lim_{n \to \infty} a_n = 0$.

5. Series diverges. Note the series is equal to $\dfrac{1}{2} \sum_{n=1}^{\infty} \dfrac{1}{n}$, and the harmonic series diverges.

> See important fact #5 in the reading section.

6. Series converges. Note that

$$1 + 2 + 3 + \sum_{n=1}^{\infty} \frac{1}{2^n} = 6 + \frac{\frac{1}{2}}{1 - \frac{1}{2}} = 7.$$

> See important fact #6 in the reading section.

7. The series converges. Note that writing

$$\frac{1}{2} + \frac{1}{6} + \frac{1}{12} + \frac{1}{20} + \frac{1}{30} + \cdots \text{ as}$$

$$1 - \frac{1}{2} + \frac{1}{2} - \frac{1}{3} + \frac{1}{3} - \frac{1}{4} + \frac{1}{4} - \frac{1}{5} + \cdots \text{ makes it}$$

easy to see that the series converges to 1.

> Looking at an algebraic expression in a different form sometimes makes a problem much easier.

*Section 7.2: Tests for Convergence

For convenience, the symbolism $\sum a_n$ will represent $\sum_{n=1}^{\infty} a_n$. Before describing the various tests used to determine convergence, the distinction between *absolute* and *conditional* convergence must be made.

Definitions:
  (1) A series $\sum a_n$ is *absolutely convergent* if the series

$$\sum |a_n| \text{ converges.}$$

  (2) If $\sum a_n$ converges, but $\sum |a_n|$ does not converge, then the series is *conditionally convergent*.

Authors' Note: It should be noted that absolute convergence necessarily implies conditional convergence:

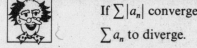

If $\sum |a_n|$ converges, then $\sum a_n$ must also. It is not possible for $\sum |a_n|$ to converge *and* for $\sum a_n$ to diverge.

Example: The alternating harmonic series $\sum (-1)^n \frac{1}{n}$ is an example of a conditionally convergent series since $\sum \frac{1}{n}$ diverges but $\sum (-1)^n \frac{1}{n}$ converges. (The proof can be seen in the test for alternating series coming up.)

Tests for Convergence of Series

1. **Alternating Series**
   A series $\sum a_n$ which is alternating in sign will converge if:

      (1) $|a_{n+1}| \leq |a_n|$

   and   (2) $\lim\limits_{n \to \infty} a_n = 0$.

Authors' Note: This test is easy to use since it requires only that (1) consecutive terms are compared and (2) the limit of the nth term is 0. Remember that *both* conditions must hold.

Example: Test the convergence of $\sum (-1)^n \frac{1}{n}$.

   This series converges since:

   (1) $\left| \dfrac{1}{n+1} \right| \leq \left| \dfrac{1}{n} \right|$ and

   (2) $\lim\limits_{n \to \infty} (-1)^n \dfrac{1}{n} = 0$.

There are several other tests used for *positive series* and they are listed below. A choice must be made as to which test is applicable and helpful. The examples listed should give hints on which to use.

2. **The Comparison Test**
   (a) If $\sum c_n$ is a positive series that converges, and $\sum a_n$ is a positive series such that eventually $a_n \leq c_n$ for all n, *then* $\sum a_n$ will also converge.
   (b) If $\sum d_n$ is a positive series that diverges, and $\sum a_n$ is a positive series such that eventually $a_n \geq d_n$ for all n, *then* $\sum a_n$ will also diverge.

Authors' Note: For the comparison test to work, it is only necessary for the terms to be eventually less than the convergent terms (or greater than the divergent terms) for all terms after a certain value for n. A finite number of terms which are not less than the corresponding terms in the convergent series will not affect the convergence.

Example: Test for convergence of $\sum \dfrac{1}{n!} = \dfrac{1}{1} + \dfrac{1}{2} + \dfrac{1}{6} + \dfrac{1}{24} + \cdots$.

   Compare the series to the geometric series $\left(\dfrac{1}{2} \right)^{n-1} = 1 + \dfrac{1}{2} + \dfrac{1}{4} + \dfrac{1}{8} + \cdots$. It can be seen that $\sum \dfrac{1}{n!}$ is term by term less than or equal to $\sum \dfrac{1}{2^{n-1}}$ and so it also must converge since the geometric series $\sum \dfrac{1}{2^{n-1}}$ converges.

Example: Test for convergence of $\sum \dfrac{1}{\ln{(n)}}$.

Compare the series to $\sum \dfrac{1}{n}$. It can be seen that since $n > \ln{(n)}$ then the series is always greater than the harmonic series.

Authors' Note: The comparison will only be conclusive if you show that a series is:

        (1) less than a convergent *or*
        (2) greater than a divergent.
Proving a series is greater than a convergent series for example is inconclusive.

3. **Integral Test**
   If $f(x)$ is a positive, continuous, decreasing function, the series $\sum a_n$ formed by $a_n = f(n)$ will converge if the integral $\displaystyle\int_1^\infty f(x)\,dx$ converges. In other words $\sum a_n$ and $\displaystyle\int_1^\infty f(x)\,dx$ behave in the same way, either both convergent or both divergent.

Example: The harmonic series (which we already know diverges) can be easily shown to diverge by the integral test.

$$\int_1^\infty \frac{1}{x}\,dx = \ln x \Big|_1^\infty = \ln \infty = \infty\,; \text{ therefore, the series diverges.}$$

4. **Ratio Test**
   Find the limit: $L = \lim\limits_{n \to \infty} \left| \dfrac{a_n + 1}{a_n} \right|$
       (1) If $L < 1$, then the series converges.
       (2) If $L > 1$, then the series diverges (including $L = \infty$).
       (3) If $L = 1$, the test is inconclusive and another test must be used.

Example: Examine the convergence of $\sum \dfrac{1}{n!}$.

We have $\lim\limits_{n \to \infty} \left| \dfrac{a_{n+1}}{a_n} \right| = \lim\limits_{n \to \infty} \dfrac{\dfrac{1}{(n+1)!}}{\dfrac{1}{n!}} = \lim\limits_{n \to \infty} \dfrac{n!}{(n+1)!} = \lim\limits_{n \to \infty} \dfrac{1}{n+1} = 0.$ Since $L = 0 < 1$, the series converges.

Authors' Note: Notice that the harmonic series and the series $\sum \dfrac{1}{n^2}$ are inconclusive under this test. For the harmonic series

$$\lim_{n \to \infty} \frac{\dfrac{1}{n+1}}{\dfrac{1}{n}} = \lim_{n \to \infty} \frac{n}{n+1} = 1.$$

For the series $\dfrac{1}{n^2}$, we have

$$\lim_{n \to \infty} \dfrac{\dfrac{1}{(n+1)^2}}{\dfrac{1}{n^2}} = \lim_{n \to \infty} \dfrac{n^2}{n^2 + 2n + 1} = 1.$$

5. *p*-series

A series of the form $\sum \dfrac{1}{n^p}$ is called a *p*-series. It will converge if and only if $p > 1$.

Authors' Note: If $p = 1$ the *p*-series is of course the harmonic series. Convergence for other *p*-series can be tested by the integral test, and the following results can be found. They are useful to remember as common series to be used for comparison.

$$\sum \dfrac{1}{\sqrt{n}} \text{ diverges} \quad (p = 1/2)$$

$$\sum \dfrac{1}{n^{3/2}} \text{ converges} \quad (p = 3/2)$$

$$\sum \dfrac{1}{n^2} \text{ converges} \quad (p = 2)$$

6. Quotient Test

If $\sum a_n$ and $\sum b_n$ are two positive series and if $\lim\limits_{n \to \infty} \dfrac{a_n}{b_n} = C$ (where $C \neq 0$), *then* the two series behave the same way; that is, if one series converges, then so does the other *and* if one series diverges, then so does the other.

Authors' Note: It is useful to learn the convergence and divergence of the series which have been mentioned in the examples for use in comparison.

Experience will teach that certain tests are more helpful for certain forms: terms that can be integrated easily suggest the integral test, factorial notation lends itself to a ratio of terms, and so on. Sometimes it is helpful to try to guess convergence or divergence first before proceeding.

Problem Set 7.2

In problems 1–9, test the series for convergence or divergence.

1. $\displaystyle\sum_{n=1}^{\infty} \dfrac{1}{2n-1}$

2. $\displaystyle\sum_{n=1}^{\infty} \dfrac{n+1}{n+2}$

3. $\displaystyle\sum_{n=1}^{\infty} \dfrac{3^n}{n!}$

4. $\displaystyle\sum_{n=1}^{\infty} \dfrac{n^2}{n^3+1}$

5. $\displaystyle\sum_{n=1}^{\infty} (-1)^n \dfrac{n^2}{n^3+1}$

6. $\displaystyle\sum_{n=1}^{\infty} \dfrac{n}{2^n}$

7. $\displaystyle\sum_{n=1}^{\infty} \dfrac{\sin(n)}{3^n}$

8. $\displaystyle\sum_{n=1}^{\infty} \dfrac{1}{\sqrt{2n^2+3}}$

9. $\displaystyle\sum_{n=2}^{\infty} \dfrac{2n}{n^2-1}$

Solutions and Comments for Problem Set 7.2

1. The series diverges since

$$\frac{1}{2n-1} > \frac{1}{2n} = \frac{1}{2} \cdot \frac{1}{n}, \text{ which diverges.}$$

> $\frac{1}{2n}$ behaves the same way as $\frac{1}{n}$ since multiplying a series by a constant has no effect on the convergence or divergence.

2. Since $\lim\limits_{n\to\infty} a_n = \lim\limits_{n\to\infty} \dfrac{n+1}{n+2} = 1$, the series diverges.

> $\lim\limits_{n\to\infty} a_n = 0$ is a necessary but not sufficient condition for convergence. If $\sum a_n$ converges, then $\lim\limits_{n\to\infty} a_n = 0$. The converse of this is not true.

3. By the ratio test,

$$\lim_{n\to\infty} \frac{|a_{n+1}|}{|a_n|} = \frac{3^{n+1}}{(n+1)!} \cdot \frac{n!}{3^n} = \lim_{n\to\infty} \frac{3}{n+1} = 0,$$

so the series converges.

> The ratio test is often useful and effective for factorial notation.

4. Using the integral test,

$$\int_1^\infty \frac{x^2\,dx}{x^3+1} = \frac{1}{3}\ln|x^3+1|\Big|_1^\infty = \infty \Rightarrow$$

series diverges.

> For the integration, let $u = x^3 + 1$; hence $du = 3x^2\,dx$.

5. Since $\lim\limits_{n\to\infty} a_n = 0$ and the series is alternating, it converges. It is conditional convergence since $\sum |a_n|$ diverges. (Compare to problem #4.)

6. Using the ratio test.

$$\lim_{n\to\infty} \frac{|a_{n+1}|}{|a_n|} = \lim_{n\to\infty} \frac{|n+1|}{|2^{n+1}|} \times \frac{|2^n|}{|n|} = \lim_{n\to\infty} \frac{n+1}{2n} = \frac{1}{2}.$$

Hence series converges.

7. Use comparison test. We have

$$\frac{\sin n}{3^n} \le \frac{1}{3^n} = \left(\frac{1}{3}\right)^n. \text{ Since } \sum\left(\frac{1}{3}\right)^n \text{ converges, so}$$

does the given series.

8. Comparison test. If $n > 1$, $\dfrac{1}{\sqrt{2n^2+3}} > \dfrac{1}{\sqrt{4n^2}} = \dfrac{1}{2n}$. Since $\sum \dfrac{1}{2n}$ diverges, so does the given series.

9. Comparison test. $\dfrac{2n}{n^2-1} > \dfrac{2n}{n^2} = \dfrac{2}{n}$. Since $\sum \dfrac{2}{n}$ diverges, so does the given series.

> This can also be done by the integral test.

*Section 7.3: Power Series

A *power series* of x is an expression of the form:

$$(1) \quad \sum_{n=0}^{\infty} a_n x^n = a_0 + a_1 x + a_2 x^2 + \cdots.$$

In general, the series will converge for certain values of x and the task is to find the set, called the interval of convergence, for which the series converges. The *interval of convergence* is the domain of the function which the power series defines.

A more generalized form of power series is

$$(2) \quad \sum_{n=0}^{\infty} a_n (x-a)^n = a_0 + a_1(x-a) + a_2(x-a)^2 + \cdots$$

(power series of $x - a$).   *(next page)*

If $x = 0$ in (1) the series converges to a_0, and if $x = a$ in (2) the series converges to a_0.

There are several important ideas concerning power series:

(1) The *radius of convergence* is the "right" endpoint of the interval of convergence.

(2) If p is the radius of convergence, then there is absolute convergence for all x such that $|x - a| < p$. The series diverges for all x such that $|x - a| > p$.

(In other words, there is absolute convergence inside the interval.)

(3) The endpoints of the interval must be checked separately for convergence.

(4) The number $x = a$ is the center of the interval.

(5) The interval of convergence can be found by using the ratio test.

Q1. Find the interval of convergence for the power series $\sum_{n=1}^{\infty} \frac{x^n}{n}$. _____

Answer: $\lim_{n \to \infty} \frac{|a_{n+1}|}{|a_n|} = \lim_{n \to \infty} \frac{|x^{n+1}|}{n+1} \times \frac{n}{|x^n|} = \lim_{n \to \infty} \frac{n}{n+1}|x| = |x|$. Hence, using the ratio test, the series converges for $|x| < 1$.

If $x = 1$, $\sum \frac{1}{n}$ diverges.

If $x = -1$, $\sum (-1)^n \frac{1}{n}$ converges.

The desired interval of convergence is thus $-1 \le x < 1$.

Functions can be expressed as power series for all x's in the interval of convergence. If $f(x)$ is such a function, then it is defined by

$$f(x) = \sum_{n=0}^{\infty} a_n(x - a)^n = a_0 + a_1(x - a) + a_2(x - a)^2 + a_3(x - a)^3 + \cdots$$

The function has the same properties as a polynomial:

(1) It is continuous within the interval of convergence.

(2) The series can be differentiated or integrated term by term, provided $f(x)$ is differentiable.

(3) The series can be added, subtracted, multiplied or divided term by term.

If you are given a certain function, the power series associated with it can be found by the *Taylor Series* expansion, which allows you to find the coefficients by taking successive derivatives.

$$f(x) = f(a) + f'(a)(x - a) + \frac{f''(a)(x - a)^2}{2!} + \frac{f'''(a)(x - a)^3}{3!} + \cdots + \frac{f^{(n)}(a)(x - a)^n}{n!} + \cdots$$

The *Maclaurin Series* is a special case of the Taylor Series when $a = 0$.

$$f(x) = f(0) + f'(0)x + \frac{f''(0)}{2!}x^2 + \frac{f'''(0)}{3!}x^3 + \cdots$$

The Taylor expansion above is expressed as an infinite series. The remaining terms after the term containing the n-th derivative can be expressed as a remainder according to Taylor's Theorem:

$$f(x) = f(a) + \sum_{k=1}^{n} f^{(n)}(a)(x - a)^n + R_n(x), \quad \text{where}$$

$$R_n(x) = \frac{1}{n!} \int_a^x (x - t)^n f^{(n+1)}(t)\, dt.$$

Another form of the remainder which is sometimes easier to estimate is Lagrange's form:

$$R_n(x) = \frac{f^{(n+1)}(c)(x-a)^{n+1}}{(n+1)!}, \quad \text{where } a < c < x.$$

The series will converge for all values of x for which the remainder goes to 0. The task then is to solve

$$\lim_{n \to \infty} R_n(x) = 0 \quad \text{for } x.$$

Example: Use the Taylor Series expansion to express $x^2 + 2$ in powers of $x - 1$.
In this situation $a = 1$.

$$f(x) = x^2 + 2 \Rightarrow f(1) = 3.$$
$$f'(x) = 2x \Rightarrow f'(1) = 2.$$
$$f''(x) = 2 \Rightarrow f''(1) = 2.$$
$$f'''(x) = 0 \Rightarrow f'''(1) = 0.$$
$$f^{(n)}(x) = 0 \text{ for } n = 3, 4, 5, \ldots.$$

Hence $x^2 + 2 = 3 + 2(x - 1) + \frac{2}{2}(x - 1)^2 = 3 + 2(x - 1) + (x - 1)^2$.

Example: Find the first four terms in the Maclaurin Series for $\cos x$.

$$f(x) = \cos x \Rightarrow f(0) = 1.$$
$$f'(x) = -\sin x \Rightarrow f'(0) = 0.$$
$$f''(x) = -\cos x \Rightarrow f''(0) = -1.$$
$$f'''(x) = \sin x \Rightarrow f'''(0) = 0.$$
$$f^{(4)}(x) = \cos x \Rightarrow f^{(4)}(0) = 1.$$
$$f^{(5)}(x) = -\sin x \Rightarrow f^{(5)}(0) = 0.$$
$$f^{(6)}(x) = -\cos x \Rightarrow f^{(6)}(0) = -1.$$

Hence, $\cos x = 1 + 0x + \dfrac{-1}{2!}x^2 + 0x^3 + \dfrac{1}{4!}x^4 + 0x^5 + \dfrac{-1}{6!}x^6 + \cdots$

$$= 1 - \frac{x^2}{2!} + \frac{x^4}{4!} - \frac{x^6}{6!} + \cdots.$$

Authors' Note: When creating power series, the selection of the number, a, must be done carefully. Remember: Certain functions are not defined over the entire set of real numbers. For instance, $\ln x$ is not defined at $x = 0$ and $\tan x$ is not defined at $x = \pi/2$.

Problem Set 7.3

In problems 1–4, find the interval of convergence for the power series:

1. $\sum \dfrac{nx^n}{3^n}$

2. $\sum (-2)^n(n + 1)(x - 1)^n$

3. $\sum \dfrac{x^n}{n!}$

4. $\sum \dfrac{\ln n}{n} x^n$

5. (a) Write the Maclaurin Series for $f(x) = e^x$.
   (b) Calculate e to four decimal places.
   (c) Estimate the error if e is approximated by the series if $n = 10$.

6. (a) Write the Maclaurin Series for $\sin x$.
   (b) Show that $\dfrac{d(\sin x)}{dx} = \cos x$ by differentiating term by term.

7. (a) Write a power series for $f(x) = \sqrt{x}$, $a = 9$ (first three terms).

   (b) Use this to estimate $\sqrt{10}$.

8. Write a power series for $f(x) = \ln x$, $a = 1$ and use this to find $\ln(1.2)$.

Solutions and Comments for Problem Set 7.3

1. $\lim\limits_{n \to \infty} \left| \dfrac{(n+1)x^{n+1}}{3^{n+1}} \right| \times \left| \dfrac{3^n}{nx^n} \right| = \lim\limits_{n \to \infty} \dfrac{n+1}{3n} |x|$

   $\qquad = \dfrac{|x|}{3} < 1$ for $|x| < 3$.

   If $x = 3$, $\sum n$ diverges. If $x = -3$, $\sum (-1)^n n$ diverges. Hence the interval of convergence is $-3 < x < 3$.

> The ratio test gives the interval. Remember to check the endpoints for convergence.

2. $\lim\limits_{n \to \infty} \dfrac{|a_{n+1}|}{|a_n|} = \lim\limits_{n \to \infty} \dfrac{|(-2)^{n+1}(n+2)(x-1)^{n+1}|}{|(-2)^n(n+1)(x-1)^n|} =$

   $\lim\limits_{n \to \infty} \left| (-2)\dfrac{n+2}{n+1}(x-1) \right| = 2|x-1| < 1$ if

   $|x - 1| < 1/2$. Interval of convergence is $\frac{1}{2} < x < \frac{3}{2}$. Series diverges if $x = \frac{1}{2}$ or $x = \frac{3}{2}$.

3. This is a positive series. $\lim\limits_{n \to \infty} \dfrac{(x^{n+1})n!}{(n+1)!(x^n)} =$

   $\lim\limits_{n \to \infty} \dfrac{1}{n+1}|x| = 0$. Hence series converges for all x.

4. $\lim\limits_{n \to \infty} \dfrac{\ln(n+1)}{\ln(n)} \times \dfrac{n}{n+1} \times \dfrac{|x^{n+1}|}{|x^n|} = |x|$. Series

   converges for $|x| < 1$. If $x = 1$, $\sum \dfrac{\ln(n)}{n}$ diverges

   by the integral test. If $x = -1$, $\sum (-1)^n \dfrac{\ln n}{n}$

   converges by the alternating series test

   $\left[\lim\limits_{n \to \infty} \dfrac{\ln n}{n} = 0 \text{ by L'Hopital's Rule} \right]$. Hence

   interval of convergence is $-1 \leq x < 1$.

5. (a) $e^x = e^0 + e^0 x + \dfrac{e^0 x^2}{2!} + \dfrac{e^0 x^3}{3!} + \cdots =$

   $1 + x + \dfrac{x^2}{2} + \dfrac{x^3}{3!} + \cdots + \dfrac{x^n}{n!} + \cdots$.

   (b) $1 + 1 + \dfrac{1}{2} + \dfrac{1}{6} + \dfrac{1}{24} + \dfrac{1}{120} + \dfrac{1}{720} + \dfrac{1}{5040} +$

   $\cdots = 2.7183$.

(c) $R_{10}(x) = \dfrac{1}{11!}e^c x^{11}$; $R_{10}(1) = \dfrac{1}{11!}e^c \leq \dfrac{1}{11!}e$

   $= \dfrac{e}{39,916,800}$ since $0 < c < 1$.

> Note that differentiating the Maclaurin Series for e^x merely reproduces the same series. Remember that $\dfrac{d}{dx}e^x = e^x$.

6. (a) $\sin x = x - \dfrac{x^3}{3!} + \dfrac{x^5}{5!} - \dfrac{x^7}{7!} + \cdots +$

   $\dfrac{(-1)^{n-1}x^{2n-1}}{(2n-1)!}$.

   (b) Differentiating the series in (a) term by term produces $1 - \dfrac{x^2}{2} + \dfrac{x^4}{4!} - \dfrac{x^6}{6!} + \cdots$, which is the Maclaurin Series for $\cos x$.

7. (a) $f(x) = x^{1/2} \Rightarrow f(9) = 3$; $f'(x) = \frac{1}{2}x^{-1/2} \Rightarrow$ $f'(9) = \frac{1}{6}$; $f''(x) = -\frac{1}{4}x^{-3/2} \Rightarrow f''(9) = -1/108$. Hence $f(x) = x^{1/2} =$

   $3 + \frac{1}{6}(x - 9) - \frac{1}{108} \times \dfrac{(x - 9)^2}{2!} + \cdots$.

   (b) $\sqrt{10} = f(10) = 3 + \frac{1}{6} - \frac{1}{108} \times \frac{1}{2} + \cdots$, or approximately 3.162037.

> Calculator value: 3.1622776.

8. $f(x) = \ln x \Rightarrow f(1) = 0$; $f'(x) = \frac{1}{x} \Rightarrow$ $f'(1) = 1$; $f''(x) = -x^{-2} \Rightarrow f''(1) = -1$; $f'''(x) = 2x^{-3} \Rightarrow f'''(1) = 2$.

   $\ln x = (x - 1) - \dfrac{(x - 1)^2}{2} + \dfrac{(x - 1)^3}{3} -$

   $\dfrac{(x - 1)^4}{4} + \cdots$.

   $\ln(1.2) = 0.2 - \dfrac{(0.2)^2}{2} + \dfrac{(0.2)^3}{3} - \dfrac{(0.2)^4}{4} +$

   $\cdots$, or about 0.1823.

> Calculator value: 0.182321557.

*Section 7.4: Multiple Choice Questions for Chapter 7

1. $\displaystyle\sum_{k=0}^{\infty} \frac{3}{10^{k+1}} =$
   (A) 3/10
   (B) 1/3
   (C) 3/7
   (D) 20/33
   (E) 6/11

2. The first six terms in the sequence defined recursively by $a_1 = 1, a_2 = 1, a_n = a_{n-1} + a_{n-2}$ are
   (A) $\{1, 1, 2, 3, 5, 8\}$
   (B) $\{1, 1, 2, 4, 6, 10\}$
   (C) $\{1, 1, 4, 5, 9, 14\}$
   (D) $\{1, 1, 0, 1, 0, 1\}$
   (E) None of the answer choice is correct.

3. A formula for the n-th term of the series
   $1 - \dfrac{1}{3} + \dfrac{1}{5} - \dfrac{1}{7} + \dfrac{1}{9} - \cdots$ is
   (A) $\dfrac{1}{2n + 1}$
   (B) $\dfrac{(-1)^n}{2n - 1}$
   (C) $\dfrac{(-1)^{n+1}}{n + 2}$
   (D) $\dfrac{(-1)^{n+1}}{4n - 1}$
   (E) $\dfrac{(-1)^{n+1}}{2n - 1}$

4. Which of the following series converge(s)?
   (I) $\displaystyle\sum_{n=1}^{\infty} \frac{\cos n}{2^n}$
   (II) $\displaystyle\sum_{n=1}^{\infty} \frac{1}{(\sqrt{5} + 1)^n}$
   (III) $\displaystyle\sum_{n=1}^{\infty} \frac{2 + \sin n}{n}$
   (A) I only
   (B) II only
   (C) II and III
   (D) I and II
   (E) I, II and III

5. Consider the series $\displaystyle\sum_{n=1}^{\infty} \frac{n^n}{n!}$. Find $\displaystyle\lim_{n\to\infty} \frac{a_{n+1}}{a_n}$.
   (A) 1
   (B) 3/2

   (C) e
   (D) 1/2
   (E) 2/3

6. The integral test confirms that the series $\displaystyle\sum_{n=1}^{\infty} \frac{e^{1/n}}{n^2}$ converges. What is $\displaystyle\int_{1}^{\infty} \frac{e^{1/x}}{x^2}\,dx$?
   (A) $e + 1$
   (B) $e - 1$
   (C) $\sqrt{e} - 1$
   (D) e^2
   (E) $e^{1/4} - 1$

7. What is the radius of convergence for the power series $\displaystyle\sum_{n=0}^{\infty} \left(-\frac{x}{3}\right)^n$?
   (A) 1/3
   (B) 1/2
   (C) 1
   (D) 3
   (E) 6

8. The series $\displaystyle\sum_{n=0}^{\infty} \frac{x^n}{n!}$
   (A) converges only if $0 < x < \frac{1}{2}$
   (B) converges only if $-1 < x < 1$
   (C) converges only if $x = 0$
   (D) converges for all values of x
   (E) diverges

9. The first three terms in the Maclaurin Series for $\ln(1 + x)$ are
   (A) $1 - \dfrac{x}{2} + \dfrac{x^2}{3}$
   (B) $x + \dfrac{x^2}{2} + \dfrac{x^3}{3}$
   (C) $x - \dfrac{x^2}{2} + \dfrac{x^3}{3}$
   (D) $1 - \dfrac{x^3}{3} + \dfrac{x^5}{5}$
   (E) $1 + \dfrac{x^3}{3} + \dfrac{x^5}{5}$

10. The first three terms in the Maclaurin Series for $\dfrac{e^x + e^{-x}}{2}$ are
    (A) $1 + \dfrac{x^2}{2} + \dfrac{x^4}{4}$

(B) $1 + \dfrac{x^2}{2!} + \dfrac{x^4}{4!}$

(C) $x + \dfrac{x^3}{3} + \dfrac{x^5}{5}$

(D) $x + \dfrac{x^3}{3!} + \dfrac{x^5}{5!}$

(E) $1 + \dfrac{x^2}{2!} + \dfrac{x^3}{3!}$

11. *How many* of the series shown diverge?

(i) $\displaystyle\sum_{n=1}^{\infty} \left(\dfrac{1}{2}\right)^n$

(ii) $\displaystyle\sum_{n=1}^{\infty} \dfrac{1}{n!}$

(iii) $\displaystyle\sum_{n=1}^{\infty} \dfrac{1}{n}$

(iv) $\displaystyle\sum_{n=1}^{\infty} \dfrac{1}{n^{3/2}}$

(v) $\displaystyle\sum_{n=1}^{\infty} \dfrac{1}{n^2}$

(A) 1

(B) 2
(C) 3
(D) 4
(E) 5

12. *How many* of the series shown converge?

(i) $\displaystyle\sum_{n=1}^{\infty} \dfrac{1}{\ln n}$

(ii) $\displaystyle\sum_{n=1}^{\infty} \dfrac{1}{n}$

(iii) $\displaystyle\sum_{n=1}^{\infty} \dfrac{1}{e^n}$

(iv) $\displaystyle\sum_{n=1}^{\infty} \dfrac{1}{\sqrt{n}}$

(v) $\displaystyle\sum_{n=1}^{\infty} n$

(A) 1
(B) 2
(C) 3
(D) 4
(E) 5

Solutions and Comments for Multiple Choice Questions

1. **(B)** $\dfrac{3}{10}\left[1 + \dfrac{1}{10} + \dfrac{1}{10^2} + \cdots\right] = \dfrac{3}{10}\left(\dfrac{1}{1 - \frac{1}{10}}\right)$

$= \dfrac{1}{3}.$

Geometric series.

2. **(A)** Each term is the sum of the two preceding terms.

This is the famous Fibonacci sequence.

3. **(E)**

Careless response would be (B). In (B), the odd-numbered terms would be negative.

4. **(D)** Comparison to $\sum \dfrac{1}{2^n}$, which converges, shows that (I) and (II) converge. Terms in (III)

are equal to or greater than terms in $\sum\dfrac{1}{n}$, which diverges.

5. **(C)** $\dfrac{a_{n+1}}{a_n} = \dfrac{(n+1)^n}{n^n} = \left(\dfrac{n+1}{n}\right)^n = \left(1 + \dfrac{1}{n}\right)^n.$

$\displaystyle\lim_{n\to\infty}\left(1 + \dfrac{1}{n}\right)^n = e.$

6. **(B)** $\displaystyle\lim_{b\to\infty} \int_1^b \dfrac{e^{1/x}}{x^2}\,dx = \lim_{b\to\infty} \left(-e^{1/x}\right)\Big|_1^b$

$= \displaystyle\lim_{b\to\infty} \left[-e^{1/b} - (-e)\right]$

$= -1 + e$

7. **(D)** Using the ratio test,

$\displaystyle\lim_{n\to\infty}\left|\dfrac{a_{n+1}}{a_n}\right| = \lim_{n\to\infty}\dfrac{|-x|}{3} = \dfrac{|x|}{3}.$

$\dfrac{|x|}{3} < 1$ when $|x| < 3.$

8. **(D)** $\displaystyle\lim_{n\to\infty}\left|\dfrac{a_{n+1}}{a_n}\right| = \lim_{n\to\infty}\left|\dfrac{x}{n+1}\right| = 0.$

Since $0 < 1$, it follows that the series converges for all x.

9. **(C)** $f(x) = \ln(1 + x) \Rightarrow f(0) = 0.$
$f'(x) = (1 + x)^{-1} \Rightarrow f'(0) = 1.$
$f''(x) = -(1 + x)^{-2} \Rightarrow f''(0) = -1.$
$f'''(x) = 2(1 + x)^{-3} \Rightarrow f'''(0) = 2.$

Series is $0 + 1 \times x + \dfrac{(-1)}{2!}x^2 + \dfrac{2}{3!}x^3 + \cdots.$

10. **(B)** $f(x) = f''(x) = f^{(4)}(x) \Rightarrow$
$f(0) = f''(0) = f^{(4)}(0) = 1.$
$f'(x) = f'''(x) = f^{(5)}(x) \Rightarrow$
$f'(0) = f'''(0) = f^{(5)}(0) = 0.$

$$\boxed{\dfrac{e^x + e^{-x}}{2} = \cosh x.}$$

11. **(A)** (iii) is the only divergent series. It is the famous harmonic series.

12. **(A)** (iii) is the only convergent series.

$$\boxed{\begin{array}{l} \text{The following series are frequently referred} \\ \text{to for comparison purposes:} \\ \text{Divergent series: } \sum \dfrac{1}{\ln n}, \sum \dfrac{1}{n}, \sum \dfrac{1}{\sqrt{n}}. \\ \text{Convergent series: } \sum \dfrac{1}{n!}, \sum \left(\dfrac{1}{2}\right)^n, \sum \dfrac{1}{n^{3/2}}, \\ \sum \dfrac{1}{n^2}. \end{array}}$$

Chapter 8

DIFFERENTIAL EQUATIONS

*Section 8.1: Introduction to Differential Equations

A *differential equation* is an equation that contains one or more derivatives or differentials. Differential equations are classified according to *order* and *degree*. The *order* of a differential equation is the order of the highest-order derivative in the equation. The *degree* is the exponent of the highest power of the highest order derivative. In general, the equation must be a polynomial in terms of a dependent variable and its derivatives.

Example: $\left(\dfrac{d^2 y}{dx^2}\right)^3 + x\left(\dfrac{dy}{dx}\right)^2 + \dfrac{y^2}{x - 1} = \sin x$ is a differential equation of order 2 and degree 3.

A *solution* to a differential equation is an equation without derivatives which satisfies the given differential equation. That is, if you differentiate the solution and substitute the appropriate derivatives into the equation, the resulting equation (statement) will be true.

Example: A solution of the differential equation $x\dfrac{dy}{dx} + 1 = 0$ would be $y = -\ln x + 3$. Note that $\dfrac{dy}{dx} = -\dfrac{1}{x}$

and that this is a solution to the given equation since $x\left(-\dfrac{1}{x}\right) + 1 = 0$. Note also that this solution is not unique.

Differential equations involving one independent variable are called *ordinary* differential equations and will be the subject of this chapter.

*Section 8.2: First Order Differential Equations—
Variables Separable

It is sometimes possible to separate variables and write a differential equation in the form

$$f(y)\,dy + g(x)\,dx = 0.$$

If this can be done, then the general solution can be found by integrating: *(next page)*

$$\int f(y)\,dy + \int g(x)\,dx = C.$$

Q1. Solve the differential equation $x^2 dy + (y - 3)\,dx = 0.$ _____

Answer: $x^2\,dy = (3 - y)\,dx \Rightarrow \dfrac{dy}{3 - y} = \dfrac{dx}{x^2}$

$$\Rightarrow \int \cdot \frac{dy}{3 - y} = \int \frac{dx}{x^2}$$

$$\Rightarrow -\ln|3 - y| = -x^{-1} + C, \text{ or } -\ln|3 - y| + x^{-1} = C.$$

Q2. Solve $\dfrac{dy}{dx} = \dfrac{-x^2}{3y}.$ _____

Answer: $3y\,dy = -x^2\,dx \Rightarrow \int 3y\,dy = \int -x^2\,dx$

$$\Rightarrow \frac{3y^2}{2} = -\frac{x^3}{3} + C, \text{ or } \frac{3y^2}{2} + \frac{x^3}{3} = C.$$

Problem Set 8.2

Solve the indicated differential equations.

1. $(x^3 - 1)\,dy = x^2(y - 1)\,dx$

2. $\sin x\,dx - \cos y\,dy = 0$

3. $\dfrac{dy}{dx} = e^{y-x}$

4. $\sqrt{xy}\dfrac{dy}{dx} = 2$

5. $\ln x \dfrac{dy}{dx} = \dfrac{1}{x}$

6. $y\,dx - x\,dy = 0$

7. $\dfrac{dy}{dx} = -\dfrac{2x}{y}$

Solutions and Comments for Problem Set 8.2

1. $\displaystyle\int \frac{dy}{y - 1} = \int \frac{x^2}{x^3 - 1}\,dx \Rightarrow \ln|y - 1|$

$$= \frac{1}{3}\ln|x^3 - 1| + C_1 \Rightarrow 3\ln|y - 1|$$

$$= \ln|x^3 - 1| + C.$$

2. $\displaystyle\int \sin x\,dx = \int \cos y\,dy \Rightarrow -\cos x$

$$= \sin y + C_1 \Rightarrow C = \sin y + \cos x.$$

3. $\dfrac{dy}{dx} = \dfrac{e^y}{e^x} \Rightarrow e^{-y}\,dy = e^{-x}\,dx.$ Integrating, we obtain $-e^{-y} = -e^{-x} + C.$

Many of the answers can be easily checked by differentiation. For instance in problem 2, $\dfrac{d}{dx}(\sin y + \cos x) = \dfrac{d}{dx}(C) \Rightarrow \cos y \dfrac{dy}{dx} - \sin x = 0 \Rightarrow \cos y\,dy - \sin x\,dx = 0 \Rightarrow \sin x\,dx - \cos y\,dy = 0.$ In problem 3, $\dfrac{d}{dx}(e^{-x} - e^{-y}) = \dfrac{d}{dx}(C) \Rightarrow -e^{-x} + e^{-y}\dfrac{dy}{dx} = 0 \Rightarrow \dfrac{dy}{dx} = \dfrac{e^{-x}}{e^{-y}} = e^{y-x}.$

4. $y^{1/2} \, dy = 2x^{-1/2} \, dx \Rightarrow \int \sqrt{y} \, dy = \int \dfrac{2 \, dx}{\sqrt{x}}$

$\Rightarrow 2/3 y^{3/2} = 4x^{1/2} + C.$

6. $\dfrac{dx}{x} = \dfrac{dy}{y} \Rightarrow \ln |x| = \ln |y| + C.$

7. $2x \, dx + y \, dy = 0 \Rightarrow x^2 + \dfrac{y^2}{2} = C.$

5. $dy = \dfrac{dx}{x \ln x}$. Letting $u = \ln x$ and $du = \dfrac{1}{x} dx$, we

have $\int dy = \int \dfrac{1}{u} du \Rightarrow y = \ln (\ln x) + C.$

> The solution represents a family of ellipses.

*Section 8.3: First Order Homogeneous

A *homogeneous differential equation* is one that can be put in the form

$$(1) \quad \dfrac{dy}{dx} = f\left(\dfrac{y}{x}\right).$$

Authors' Note: Many times the differential equation will be expressed in the form

$$(2) \quad f(x, y) \, dx + g(x, y) \, dy = 0$$

where f and g are homogeneous functions of the same degree. Remember that a homogeneous function of degree n is a function with the property $f(kx, ky) = k^n f(x, y)$. A homogeneous differential equation can be recognized by checking to see if the degrees of f and g in form (2) are the same.

A homogeneous differential equation can be solved by making a substitution $v = \dfrac{y}{x}$ and then separating the variables v and x. Note that $v = \dfrac{y}{x} \Rightarrow y = vx$ and $\dfrac{dy}{dx} = v + x\dfrac{dv}{dx}$. The equation $\dfrac{dy}{dx} = F(v)$ thus becomes

$v + x\dfrac{dv}{dx} = F(v) \Rightarrow v \, dx + x \, dv = F(v) \, dx \Rightarrow \dfrac{dx}{x} + \dfrac{dv}{v - F(v)} = 0$ and the variables are separated. The final

solution is obtained by replacing v by $\dfrac{y}{x}$.

Example: Consider the differential equation $(x + y) \, dy - (x - y) \, dx = 0$. Letting $f(x, y) = x - y$ and $g(x, y) = x + y$, note that

$f(kx, ky) = kx - ky = k(x - y) = kf(x, y)$ and, in a similar manner,

$g(kx, ky) = kg(x, y)$. Hence f and g are homogeneous equations of degree 1.

From the given equation, $\dfrac{dy}{dx} = \dfrac{x - y}{x + y} = \dfrac{1 - \dfrac{y}{x}}{1 + \dfrac{y}{x}} = \dfrac{1 - v}{1 + v}$, where $v = \dfrac{y}{x}$. Hence $\dfrac{dy}{dx} = F(v) = \dfrac{1 - v}{1 + v}$

and the general solution is

$$\int \dfrac{dx}{x} + \int \dfrac{dv}{v - \frac{1-v}{1+v}} = \int \dfrac{dx}{x} + \int \dfrac{dv(1 + v)}{v^2 + 2v - 1}.$$

Letting $u = v^2 + 2v - 1$, the second integral has the form $\dfrac{1}{2} \int \dfrac{du}{u}$. Hence, the solution is $\ln x +$

$\dfrac{1}{2} \ln |v^2 + 2v - 1| + C = \ln x + \dfrac{1}{2} \ln \left| \dfrac{y^2}{x^2} + \dfrac{2y}{x} - 1 \right| + C.$

Problem Set 8.3

Solve the following homogeneous differential equations.

1. $xy\,dy = (x^2 - y^2)\,dx$

2. $\dfrac{dy}{dx} = \dfrac{y^3 + x^3}{xy^2}$

3. $(xe^{y/x} - y)\,dx = -x\,dy$

4. $x^2\,dy + (y^2 - xy)\,dx = 0$

5. $(x^2 + y^2)\,dy - y^2\,dx = 0$

Solutions and Comments for Problem Set 8.3

1. $\dfrac{dy}{dx} = v + x\dfrac{dv}{dx} = \dfrac{x^2 - y^2}{xy} = \dfrac{x}{y} - \dfrac{y}{x} = \dfrac{1}{v} - v =$

$\dfrac{1 - v^2}{v}$. Separating, $v + x\dfrac{dv}{dx} = \dfrac{1 - v^2}{v} \Rightarrow$

$v^2 + vx\dfrac{dv}{dx} = 1 - v^2 \Rightarrow 2v^2 - 1 = -xv\dfrac{dv}{dx} \Rightarrow$

$xv\dfrac{dv}{dx} = 1 - 2v^2 \Rightarrow \dfrac{v\,dv}{1 - 2v^2} = \dfrac{dx}{x} \Rightarrow$

$-\tfrac{1}{4}\ln|1 - 2v^2| = \ln|x| + C \Rightarrow$

$-\tfrac{1}{4}\ln\left|1 - \dfrac{2y^2}{x^2}\right| = \ln|x| + C \Rightarrow$

$-\tfrac{1}{4}\ln\left|\dfrac{x^2 - 2y^2}{x^2}\right| = \ln|x| + C \Rightarrow$

$-\tfrac{1}{4}\ln|x^2 - 2y^2| + \tfrac{1}{4}\ln x^2 - \ln|x| = C \Rightarrow$

$-\tfrac{1}{4}\ln|x^2 - 2y^2| - \tfrac{1}{2}\ln|x| = C.$

Final forms of the answer may differ as a result of using various rules for simplifying logarithms.

---

2. $\dfrac{dy}{dx} = \dfrac{y^3 + x^3}{xy^2} = \dfrac{y^3}{xy^2} + \dfrac{x^3}{xy^2} = \dfrac{y}{x} + \dfrac{x^2}{y^2} = v + \dfrac{1}{v^2}$

$= \dfrac{v^3 + 1}{v^2} = v + x\dfrac{dv}{dx} \Rightarrow v^3 + 1 = v^3 + v^2 x\dfrac{dv}{dx} \Rightarrow$

$dx = v^2 x\,dv \Rightarrow \dfrac{dx}{x} = v^2\,dv.$ Integrating,

$\ln|x| = \dfrac{v^3}{3} + C \Rightarrow \ln|x| = \dfrac{y^3}{3x^3} + C.$

3. Starting with the formula, $\dfrac{dx}{x} + \dfrac{dv}{v - F(v)} = 0$

and using $\dfrac{dy}{dx} = -(e^v - v) = v - e^v$, we have

$\dfrac{dx}{x} + \dfrac{dv}{e^v} = 0.$ Then $\ln|x| + (-e^{-v}) = C$ or $\ln|x|$

$= e^{-y/x} + C.$

4. Dividing by x^2 and solving for $\dfrac{dy}{dx}$, $\dfrac{dy}{dx} = \dfrac{y}{x} - \dfrac{y^2}{x^2}$

$= v - v^2.$ By formula, $\dfrac{dx}{x} + \dfrac{dv}{v - (v - v^2)} = 0$ or

$\dfrac{dx}{x} + \dfrac{dv}{v^2} = 0 \Rightarrow \ln|x| - \dfrac{1}{v} = C_1$ or $\dfrac{x}{y} = \ln|x| + C.$

5. $\dfrac{dy}{dx} = \dfrac{y^2}{x^2 + y^2} = \dfrac{(y/x)^2}{1 + (y/x)^2} = \dfrac{v^2}{v^2 + 1}.$ By formula, $\dfrac{dx}{x} + \dfrac{(v^2 + 1)\,dv}{v(v^2 - v + 1)} = 0.$ Using partial fractions to integrate $\dfrac{dx}{x} + \dfrac{dv}{v} + \dfrac{dv}{v^2 - v + 1} = 0$, we get $\ln|x| + \ln|v| + \left(\dfrac{2}{\sqrt{3}}\right)\tan^{-1}\dfrac{(2v - 1)}{\sqrt{3}} = C \Rightarrow$ $\ln|y| + \left(\dfrac{2}{\sqrt{3}}\right)\tan^{-1}\left(\dfrac{(2y - x)}{\sqrt{3}x}\right) = C.$

*Section 8.4: First Order Linear Differential Equations

A *first order linear differential equation* is one that can be written in the form

$$(1) \quad \dfrac{dy}{dx} + P(x)y = Q(x). \quad \textit{(next page)}$$

To solve this differential equation, we use the function

$$p(x) = e^{\int P(x)\,dx}, \quad \text{which is called an } integrating\ factor.$$

The solution to the differential equation is

$$y = \frac{1}{p(x)}\left[\int p(x)Q(x)\,dx + C\right].$$

Example: Solve $\dfrac{dy}{dx} - y = e^{-x}$.

In this situation, $P(x) = -1$, $Q(x) = e^{-x}$, and $p(x) = e^{\int -dx} = e^{-x}$.

Hence $y = \dfrac{1}{e^{-x}}\int e^{-x}e^{-x}\,dx = \dfrac{1}{e^{-x}}\int e^{-2x}\,dx = \dfrac{1}{e^{-x}}\left(-\dfrac{1}{2}e^{-2x} + C\right) = -\dfrac{1}{2}e^{-x} + Ce^{x}$.

The following procedure is often useful:

(a) Write the differential equation in form (1) so that P and Q can be easily identified.

(b) Find $p(x) = e^{\int P(x)\,dx}$.

(c) Write $p(x)y = \displaystyle\int p(x)Q(x)\,dx + C$.

(d) Divide by $p(x)$ to obtain y.

Example: Solve $x\dfrac{dy}{dx} + y = x^{2}.\ \left[\text{Write as } \dfrac{dy}{dx} + \dfrac{1}{x}y = x.\right]$

In this case, $P(x) = \dfrac{1}{x}$, $Q(x) = x$, and $p(x) = e^{\int dx/x} = e^{\ln x} = x$.

Then, $xy = \displaystyle\int xx\,dx = \int x^{2}\,dx = \dfrac{x^{3}}{3} + C \Rightarrow y = \dfrac{x^{2}}{3} + Cx^{-1}$.

Authors' Note: Here is a brief explanation as to "why" the function $p(x) = e^{\int P(x)\,dx}$ is so useful: Note that

$$\frac{dy}{dx} + P(x)y = Q(x) \Rightarrow (e^{\int P(x)\,dx})\frac{dy}{dx} + (e^{\int P(x)\,dx})P(x)y = Q(x)e^{\int P(x)\,dx}$$

$$\Rightarrow \frac{d}{dx}(ye^{\int P(x)\,dx}) = Q(x)e^{\int P(x)\,dx}$$

$$\Rightarrow ye^{\int P(x)\,dx} = \int Q(x)e^{\int P(x)\,dx}\,dx + C$$

$$\Rightarrow y \times p(x) = \int p(x)Q(x)\,dx + C.$$

Problem Set 8.4

Solve the following differential equations.

1. $\dfrac{dy}{dx} + ky = 0$

2. $x\dfrac{dy}{dx} + 2y = \dfrac{\sin x}{x}$

3. $\dfrac{dy}{dx} + y = \sin x$

4. $\dfrac{dy}{dx} - 2y = e^{x}$

5. $(x + 1)^{2}\dfrac{dy}{dx} + 2(x + 1)y = x - 1$

6. $x\,dy - (2x - y)\,dx = 0$

Solutions and Comments for Problem Set 8.4

1. $P(x) = k$, $Q(x) = 0$, $p = e^{\int k\,dx} = e^{kx}$. Using the formula, $e^{kx}y = \int 0 \cdot dx \Rightarrow e^{kx} \cdot y = C \Rightarrow y = Ce^{-kx}$.

> This problem can also be solved easily by separating the variables.

> The integration of $e^x \sin x\,dx$ is done by using integration by parts twice.

2. The equation becomes $\dfrac{dy}{dx} + \dfrac{2}{x}y = \dfrac{\sin x}{x^2}$ and so $P(x) = 2/x$, $Q(x) = \sin(x)/x^2$, $p(x) = e^{\int(2/x)\,dx} = e^{2\ln x} = x^2$. Hence, $x^2 y = \int x^2 \cdot \dfrac{\sin x}{x^2}\,dx =$

$\int \sin x\,dx = -\cos x + C \Rightarrow y = \dfrac{-\cos x}{x^2} + Cx^{-2}$.

3. $P(x) = 1$, $Q(x) = \sin x$, $p = e^{\int dx} = e^x \Rightarrow e^x y = \int e^x \sin x \cdot dx \Rightarrow e^x y = \dfrac{e^x}{2}(\sin x - \cos x) + C \Rightarrow y = \frac{1}{2}(\sin x - \cos x) + Ce^{-x}$.

4. $P(x) = -2$, $Q(x) = e^x$, $p = e^{\int -2\,dx} = e^{-2x}$. $e^{-2x}y = \int e^{-2x}e^x\,dx \Rightarrow e^{-2x} \cdot y = \int e^{-x}\,dx \Rightarrow$

$e^{-2x} \cdot y = -e^{-x} + C \Rightarrow y = -\dfrac{e^{-x}}{e^{-2x}} + Ce^{2x} = -e^x + Ce^{2x}$.

5. $P(x) = \dfrac{2}{x+1}$, $Q(x) = \dfrac{x-1}{(x+1)^2}$, $p = e^{\int \frac{2}{x-1}\,dx} = e^{2 \times \ln|x+1|} = (x+1)^2$. Hence $(x+1)^2 y = \int (x+1)^2 \cdot \dfrac{x-1}{(x+1)^2}\,dx = \int (x-1)\,dx \Rightarrow$

$(x+1)^2 y = \dfrac{x^2}{2} - x + C \Rightarrow$

$y = \dfrac{1}{2}\left(\dfrac{x}{x+1}\right)^2 - \dfrac{x}{(x+1)^2} + C(x+1)^{-2}$.

6. $P(x) = 1/x$, $Q(x) = 2$, $p = e^{\int(1/x)\,dx} = e^{\ln x} = x$. Hence $x \cdot y = \int x \cdot 2 \cdot dx \Rightarrow xy = \dfrac{2x^2}{2} + C = x^2 + C \Rightarrow y = x + Cx^{-1}$.

*Section 8.5: Second Order Homogeneous Linear with Constant Coefficients

A *second order homogeneous differential equation with constant coefficients* can be written in the form

$$(1) \quad \frac{d^2y}{dx^2} + a\frac{dy}{dx} + by = 0.$$

It is helpful to use *operator* notation, letting $D = \dfrac{dy}{dx}$ and $D^2 = \dfrac{d^2y}{dx^2}$. Equation (1) can then be expressed as

$$(D^2 + aD + b)y = 0.$$

The first step in finding a solution for (1) is to solve the quadratic $D^2 + aD + b = 0$. This equation is called the *characteristic* or *homogeneous* equation and the roots are

$$r_1 = \frac{-a + \sqrt{a^2 - 4b}}{2} \quad \text{and} \quad r_2 = \frac{-a - \sqrt{a^2 - 4b}}{2}.$$

There are three possible cases to consider:
  (i) If $a^2 - 4b > 0$, the quadratic equation has two real roots and the solution to (1) is $y = C_1 e^{r_1 x} + C_2 e^{r_2 x}$ (C_1 and C_2 are constants). *(next page)*

(ii) If $a^2 - 4b = 0$, then the quadratic equation has one real root ($r = r_1 = r_2$) and the solution to (1) is $y = (C_1 x + C_2)e^{rx}$.

(iii) If $a^2 - 4b < 0$, the quadratic equation has two imaginary roots $r_1 = \alpha + \beta i$ and $r_2 = \alpha - \beta i$ and the solution is $y = e^{\alpha x}(C_1 \cos \beta x + C_2 \sin \beta x)$.

Q1. Solve: (a) $(D^2 - 3D - 4)y = 0$ _____ (b) $y'' + 2y' + 4y = 0$ _____

Answers: (a) $D^2 - 3D - 4 = 0 \Rightarrow (D - 4)(D + 1) = 0$. Hence $r_1 = 4$ and $r_2 = -1$. The solution is $y = C_1 e^{4x} + C_2 e^{-x}$.

(b) Solving $D^2 + 2D + 4 = 0$, we have $r_1 = -1 + \sqrt{3}i$ and $r_2 = -1 - \sqrt{3}i$. Hence $\alpha = -1$ and $\beta = \sqrt{3}$ and the solution is $y = e^{-x}(C_1 \cos \sqrt{3}x + C_2 \sin \sqrt{3}x)$.

Problem Set 8.5

Solve the following differential equations.

1. $(D^2 - 5D - 6)y = 0$

2. $(D^2 + 6D - 5)y = 0$

3. $y'' + 4y' + 4y = 0$

4. $y'' + y' + y = 0$

Solutions and Comments for Problem Set 8.5

1. $(D - 6)(D + 1) = 0 \Rightarrow r_1 = 6$ and $r_2 = -1$. Solution is $y = C_1 e^{6x} + C_2 e^{-x}$.

2. $r_1 = -3 + \sqrt{14}$ and $r_2 = -3 - \sqrt{14}$. Solution is $y = C_1 e^{(-3+\sqrt{14})x} + C_2 e^{(-3-\sqrt{14})x}$.

3. $(D + 2)^2 y = 0$, $r = -2$. Solution is $y = (C_1 x + C_2)e^{-2x}$.

4. $r_1 = -\dfrac{1}{2} + \dfrac{\sqrt{3}}{2}i$ and $r_2 = -\dfrac{1}{2} - \dfrac{\sqrt{3}}{2}i$. Solution is $y = e^{-x/2}\left(C_1 \cos \dfrac{\sqrt{3}}{2}x + C_2 \sin \dfrac{\sqrt{3}}{2}x\right)$.

> You should remember that differential equations of this type may or may not appear in operator form. For instance, problem #2 is expressed in operator form, but problem #3 is not. Remember also that you can check answers by differentiating them and substituting back into the original equation.

*Section 8.6: Second Order Non-Homogeneous Linear with Constant Coefficients

A second order non-homogeneous linear differential equation with constant coefficients can be written in the form:

$$(1) \quad \frac{d^2 y}{dx^2} + a\frac{dy}{dx} + by = F(x) \quad \text{or} \quad (D^2 + aD + b)y = F(x).$$

To find the solution, you first solve the characteristic equation $D^2 + aD + b = 0$ and obtain the solution $y_c = C_1 u_1 + C_2 u_2$ (see Section 8.5).
This is the general solution of the homogeneous equation. In this case u_1 and u_2 are functions of x.
The next step is to find a particular solution y_p of (1). The general solution will then be $y = y_c + y_p$.

The task of finding y_p requires further explanation and the method of undetermined coefficients. Specifically the task is to find y_p that will work in the original equation (1). To do this, we assume y_p to be the sum of appropriate forms (similar to $F(x)$) with undetermined coefficients and then solve for the coefficients.

Example: Solve $(D^2 - 3D + 2)y = 2x^2 + 6x + 2$.
Solving the homogeneous, $y_c = C_1 e^{2x} + C_2 e^x$. Next assume that the form of y_p will be similar to $F(x)$, so $y_p = ax^2 + bx + c$. This means $y_p' = 2ax + b$ and $y_p'' = 2a$. Substituting into the original,
$$2a - 3(2ax + b) + 2(ax^2 + bx + c) = 2x^2 + 6x + 2$$

or $\qquad 2ax^2 + (-6a + 2b)x + (2a - 3b + 2c) = 2x^2 + 6x + 2.$

This means $\quad a = 1, \quad -6 + 2b = 6 \Rightarrow b = 6, \quad 2 - 18 + 2c = 2 \Rightarrow c = 9.$

Therefore $\quad y_p = x^2 + 6x + 9 \quad$ and $\quad y = C_1 e^{2x} + C_2 e^x + x^2 + 6x + 9.$

Listed below are the forms to use for various possibilities for $F(x)$ when solving for y_p.
 If $F(x)$ is polynomial *use* y_p = polynomial with unknown coefficients.
 If $F(x)$ is e^{kx} *use* $y_p = ae^{kx}$.
 If $F(x)$ is $\cos kx$ or $\sin kx$ *use* $y_p = a \cos kx + b \sin kx$.
 If $F(x)$ is any linear combination of these *use* a corresponding combination.

Problem Set 8.6

Solve the following differential equations.

1. $(D^2 + 2D - 3)y = 9$

2. $D(D - 2)y = 6x^2 + 2x - 6$

3. $(D^2 - 2D - 3)y = e^x$

4. $(D^2 - 1)y = \sin x$

5. $(D^2 - 2D + 1)y = 3x$

6. $(D^2 - 3D + 2)y = e^{-x}$

7. $(D^2 - 2D - 3)y = 6x^2$

Solutions and Comments for Problem Set 8.6

1. By inspection $y = -3$ is a solution since y' and y'' will be 0 and so $y = y_c + y_p \Rightarrow y = C_1 e^{-3x} + C_2 e^x - 3$.

> The method of undetermined coefficients could be used to solve for the coefficients if this solution is not seen right away.

2. Solving the homogeneous equation, $y_c = C_1 e^0 + C_2 e^{2x} = C_1 + C_2 e^{2x}$. Assume the form $y_p = ax^3 + bx^2 + cx + d$. It follows that $y_p' = 3ax^2 + 2bx + c$ and that $y_p'' = 6ax + 2b$. Substituting

into the original, we get $(6ax + 2b) - (2)(3ax^2 + 2bx + c) = 6x^2 + 2x - 6 \Rightarrow (-6a)x^2 + (6a - 4b)x + (2b - 2c) = 6x^2 + 2x - 6$. Hence $-6a = 6$, $6a - 4b = 2$, and $2b - 2c = -6 \Rightarrow a = -1, b = -2,$ and $c = 1$. Therefore $y = y_c + y_p = C_1 + C_2 e^{2x} - x^3 - 2x^2 + x.$

> In this case, since there is no y term in $D(D - 2)y$, y_p must be assumed to be a cubic since a quadratic would produce a linear term at most. The constant d is included in C_1 in the last step.

3. $(D - 3)(D + 1)y = e^x$. Therefore, $y_c = C_1 e^{3x} + C_2 e^{-x}$. Assume $y_p = ae^x \Rightarrow y'_p = ae^x$, $y''_p = ae^x$. Solving for the coefficient, $ae^x - 2ae^x - 3ae^x = e^x \Rightarrow (a - 2a - 3a)e^x = e^x \Rightarrow -4a = 1 \Rightarrow a = -1/4$. Therefore, $y = y_c + y_p = C_1 e^{3x} + C_2 e^{-x} + -\dfrac{e^x}{4}$.

4. $(D + 1)(D - 1)y = \sin x$. Therefore, $y_c = C_1 e^x + C_2 e^{-x}$. Assume $y_p = a\cos x + b\sin x \Rightarrow y'_p = -a\sin x + b\cos x \Rightarrow y''_p = -a\cos x - b\sin x$. Solving for the coefficients, $-a\cos x - b\sin x - (a\cos x + b\sin x) = \sin x \Rightarrow -2a\cos x - 2b\sin x = \sin x \Rightarrow a = 0, b = -1/2 \Rightarrow y = C_1 e^x + C_2 e^{-x} + (-\tfrac{1}{2}\sin x)$.

5. $(D - 1)^2 = 3x$. Therefore $y_c = (C_1 x + C_2)e^x$. Assume $y_p = ax + b$. Then $y'_p = a$, $y''_p = 0$.

Therefore the original equation becomes $-2a + ax + b = 3x \Rightarrow ax + (-2a + b) = 3x \Rightarrow a = 3, b = 6 \Rightarrow y = (C_1 x + C_2)e^x + 3x + 6$.

6. $(D - 2)(D - 1) = e^{-x} \Rightarrow y_c = C_1 e^x + C_2 e^{2x}$. Assume $y_p = ae^{-x} \Rightarrow y'_p = -ae^{-x}$, $y''_p = ae^{-x}$. The original becomes $ae^{-x} - 3(-ae^{-x}) + 2ae^{-x} = e^{-x} \Rightarrow (a + 3a + 2a)e^{-x} = e^{-x} \Rightarrow 6a = 1 \Rightarrow a = 1/6$. Hence $y = C_1 e^x + C_2 e^{2x} + \tfrac{1}{6}e^{-x}$.

7. $(D - 3)(D + 1)y = 6x^2$. Therefore $y_c = C_1 e^{3x} + C_2 e^{-x}$. Assume $y_p = ax^2 + bx + c \Rightarrow y'_p = 2ax + b \Rightarrow y''_p = 2a$. Substituting into the original equation, $2a - 2(2ax + b) - 3(ax^2 + bx + c) = 6x^2 \Rightarrow 2a - 4ax - 2b - 3ax^2 - 3bx - 3c = 6x^2 \Rightarrow -3ax^2 + (-4a - 3b)x + (2a - 2b - 3c) = 6x^2 \Rightarrow a = -2, b = 8/3, c = -28/9$. Therefore $y = C_1 e^{3x} + C_2 e^{-x} - 2x^2 + \tfrac{8}{3}x - 28/9$.

*Section 8.7: Special Types of Differential Equations and Applications

The purpose of this section is merely to point out that some differential equations can be solved easily by special methods.
If an equation contains an exact form of a derivative of a product or a quotient, the solution is often readily forthcoming.

Example: Solve $2x\sin y\, dx - x^2 \cos y\, dy = -x\sin^2 y\, dx$.

Dividing by $\sin^2 y\, dx$ produces $\dfrac{2x\sin y - x^2 \cos y\dfrac{dy}{dx}}{\sin^2 y} = -x$.

This is $\dfrac{d}{dx}\left(\dfrac{x^2}{\sin y}\right) = -x$.

Hence the solution is $\dfrac{x^2}{\sin y} = -\dfrac{x^2}{2} + C$, or

$$\dfrac{x^2}{\sin y} + \dfrac{x^2}{2} = C.$$

Second-order equations can sometimes be written as first-order equations by using the substitutions $u = \dfrac{dy}{dx}$, $\dfrac{du}{dx} = \dfrac{d^2 y}{dx^2}$. This can often make the solution process easier.

Example: Solve $\dfrac{d^2 y}{dx^2} - \dfrac{dy}{dx} = 0$.

Using the indicated substitution, the equation becomes $u' - u = 0$. By inspection, $u = e^x$ solves this equation. Therefore $y = \int e^x\,dx = e^x + C$ is a solution to $y'' - y' = 0$.

Recall from Section 5.3 that differential equations can be used to solve practical problems involving growth and decay. Harmonic motion is another very common application, and the equation $\dfrac{d^2y}{dx^2} + k^2 y = 0$ describes this motion. The solution to this equation is

$$y = C_1 \cos kx + C_2 \sin kx.$$

Authors' Note: The characteristic equation for the harmonic motion differential equation is $D^2 + k^2 = 0$. The solutions are $0 + ki$ and $0 - ki$. From Section 8.5, the solution to the differential equation would be $y = C_1 \cos kx + C_2 \sin kx$. It is possible to write this solution in the form $y = A \cos(kx - \alpha)$. No details on the derivation of this form will be provided here, but it does have the advantage that amplitude and period are easily identifiable.

Example: Find the general solution for $y'' + 9y = 0$. Then find a particular solution if it is known that $y(0) = 5$ and $y'(0) = 8$.

The general solution is $y = C_1 \cos 3x + C_2 \sin 3x$.

Hence $y' = -3C_1 \sin 3x + 3C_2 \cos 3x$.

$y(0) = 5 \Rightarrow C_1 \cos 0 + C_2 \sin 0 = 5 \Rightarrow C_1 = 5$.

$y'(0) = 8 \Rightarrow -3C_1 \sin 0 + 3C_2 \cos 0 = 8 \Rightarrow C_2 = 8/3$.

Hence, the desired particular solution is $y = 5 \cos 3x + \dfrac{8}{3} \sin 3x$.

Problem Set 8.7

1. Solve

   (a) $x\dfrac{dy}{dx} + y = 2$

   (b) $xe^y\,dy - e^y\,dx = 14x^2\,dx$

2. Solve $y'' - 5y' = 0$.

3. The equation $y'' + 64y = 0$ represents simple harmonic motion.

   (a) Find the general solution to the equation.

   (b) Find the particular solution if $y(0) = 24$ and $y'(0) = 48$.

   (c) Find the particular solution if $y(\frac{\pi}{4}) = 7$ and $y'(\frac{\pi}{8}) = 2$.

4. A radioactive element decays with a half-life of 100 hours. How much of the original amount is left after 150 hours?

5. At a point 96 ft. above the ground a ball is thrown upward with an initial speed of 16 ft./sec. Assuming that the only force acting is that of gravity, find

   (a) how long it will take for the ball to reach the ground

   (b) the velocity with which it strikes the ground

Solutions and Comments for Problem Set 8.7

1. (a) $\dfrac{d}{dx}(xy) = 2 \Rightarrow xy = 2x + C$, or $y = 2 + Cx^{-1}$.

   (b) Dividing by $x^2\,dx$ produces $\dfrac{xe^y\dfrac{dy}{dx} - e^y}{x^2} = 14$.

   This is $\dfrac{d}{dx}\left(\dfrac{e^y}{x}\right) = 14 \Rightarrow \dfrac{e^y}{x} = 14x + C$, or $e^y = 14x^2 + Cx$.

   > Both equations contain an exact form of a derivative of a product or quotient.

2. Letting $u = y'$ and $u' = y''$ produces $u' - 5u = 0$. By inspection, $u = e^{5x}$ solves this equation. Hence a solution for the original equation is

$$y = \int e^{5x}\,dx = \frac{1}{5}e^{5x} + C.$$

3. (a) $y = C_1 \cos 8x + C_2 \sin 8x$;
   (b) $y = 24 \cos 8x + 6 \sin 8x$;
   (c) $y = 7 \cos 8x - \frac{1}{4}\sin 8x$.

$y' = -8C_1 \sin 8x + 8C_2 \cos 8x$.
(b) $y(0) = 24 \Rightarrow C_1 = 24$.
$y'(0) = 48 \Rightarrow 8C_2 = 48 \Rightarrow C_2 = 6$.
(c) $y(\frac{\pi}{4}) = 7 \Rightarrow C_1 \cos 2\pi + C_2 \sin 2\pi = 7$
$\Rightarrow C_1 = 7$.
$y'(\frac{\pi}{8}) = 2 \Rightarrow -8C_1 \sin \pi + 8C_2 \cos \pi = 2$
$\Rightarrow -8C_2 = 2 \Rightarrow C_2 = -\frac{1}{4}$.

---

4. $\dfrac{dx}{dt} = -kx \Rightarrow x = Ce^{-kt}$. When $t = 0$, $x = C$ (the original amount). When $t = 100$, we have $\frac{1}{2}C = Ce^{-100k} \Rightarrow e^{100k} = 2 \Rightarrow k = \dfrac{\ln 2}{100}$. The amount re-

maining at $t = 150$ is $x_{150} = Ce^{(-\ln 2/100)150}$ and the percent remaining is $\dfrac{x_{150}}{C} = e^{(-\ln 2/100)(150)}$, or about 35%.

This type of problem appears in Section 5.3.

---

5. $\dfrac{dv}{dt} = -32 \Rightarrow v = -32t + 16$ since the initial upward velocity is 16. If y represents distance above the ground, then $y = -16t^2 + 16t + 96 = -16(t^2 - t - 6) = -16(t - 3)(t + 2)$. $y = 0$ when $t = 3$.

   (a) 3 seconds;
   (b) At $t = 3$, $v = -32(3) + 16 = -80$.

This type of problem appears in Section 5.2. The ball is travelling at a *speed* of 80 ft./sec. when it hits the ground.

---

*Section 8.8: Multiple Choice Questions for Chapter 8

1. The solution for $\dfrac{dy}{dx} = x$ is

   (A) $\dfrac{x}{2} + C$

   (B) $x^2 + C$

   (C) $x^{1/2} + C$

   (D) $\dfrac{x^2}{2} + C$

   (E) $\frac{1}{2}\sqrt{x} + C$

2. The solution to $\dfrac{dy}{dx} = \dfrac{-x}{y}$ is

   (A) $x^2 + y^2 = C$

   (B) $y + \dfrac{x^2}{2} = C$

   (C) $y^2 = 2x + C$

   (D) $y^2 = 2x^2 + C$

   (E) $x^2 - y^2 = C$

3. Which differential equation has a solution consisting of a family of lines through the origin?

   (A) $y' = xy$
   (B) $y' = y$
   (C) $y' = x^{-1}$
   (D) $y' = 1$
   (E) $y' = yx^{-1}$

4. If the differential equation $xy' - x^2 - y = 0$ is written in the form $y' + P(x)y = Q(x)$, then

   $$\int P(x)\,dx =$$

   (A) $\frac{1}{2}x^{-1/2} + C$
   (B) $-\ln|x| + C$
   (C) $e^x + C$
   (D) $x^2 + C$
   (E) $\frac{1}{2}x^2 + C$

5. The solution to $y' - 8xy = x$ is
   (A) $\frac{1}{4} + Ce^{4x^2}$
   (B) $-\frac{1}{4}e^{-4x^2} + Ce^{4x^2}$
   (C) $-\frac{1}{8} + Ce^{4x^2}$
   (D) $-e^{-4x^2} + C$
   (E) $e^{4x^2} + Ce^{-4x^2}$

6. The general solution to $y'' - 4y' + 3y = 0$ is
   (A) $y = (C_1 x + C_2)e^{3x}$
   (B) $y = C_1 e^{-3x} + C_2 e^{-x}$
   (C) $y = e^x(C_1 \cos x + C_2 \sin x)$
   (D) $y = C_1 e^{3x} + C_2 e^x$
   (E) $y = C_1(3x) + C_2 x$

7. If $\dfrac{dy}{dx} = x^2$ and $y = 4$ when $x = 3$, then $y =$
   (A) $\dfrac{x^3}{3} - 5$
   (B) $\dfrac{x^3}{3} + 3$
   (C) $x^3 - 23$
   (D) $x + 1$
   (E) $\dfrac{x^3}{3} + 4$

8. The general solution to $\dfrac{dy}{dx} = e^{-y}$ is
   (A) $y = e^{-x} + C$
   (B) $y = \ln|x + C|$
   (C) $y = \ln|x| + C$
   (D) $y = -\ln|x| + C$
   (E) $e^y = \ln|x| + C$

9. A function which satisfies $\dfrac{dy}{dx} - y = 0$ is
   (A) $y = \sin x$
   (B) $y = \cos x$
   (C) $y = \ln x$
   (D) $y = e^x$
   (E) $y = e^{-x}$

10. If $\dfrac{dy}{dx} = \dfrac{1}{1 + x^2}$ and $y = 2$ when $x = 1$, then $y =$
    (A) $\tan^{-1} x + 1$
    (B) $\tan^{-1} x + 2$
    (C) $\tan^{-1} x + 2 - \frac{\pi}{4}$
    (D) $\tan^{-1} x + \frac{\pi}{4} + 2$
    (E) $\tan^{-1} x + \frac{\pi}{4}$

11. The general solution of $y'' - 3y' = 6$ is
    (A) $y = -2x + C$
    (B) $y = -2x$
    (C) $y = C_1 + C_2 e^{3x}$
    (D) $y = C_1 e^x + C_2 e^{3x}$
    (E) $y = C_1 + C_2 e^{3x} - 2x$

12. A particular solution of $(D^2 + 5D + 6)y = 6$ is
    (A) $y = x$
    (B) $y = 1$
    (C) $y = -1$
    (D) $y = 6x$
    (E) $y = e^{6x}$

Solutions and Comments for Multiple Choice Problems

1. **(D)** Solution is simply $\displaystyle\int x\,dx$.

2. **(A)** $y\,dy = -x\,dx \Rightarrow \dfrac{y^2}{2} = -\dfrac{x^2}{2} + C_1 \Rightarrow x^2 + y^2 = C.$

   > Solution represents a family of circles with center at $(0, 0)$.

3. **(E)** The slope of any line through the origin is merely the ratio of any non-zero y-coordinate to the corresponding x-coordinate.

4. **(B)** We have $y' + \left(-\dfrac{1}{x}\right)y = x$. Hence
   $$\int -\dfrac{1}{x}\,dx = -\ln|x| + C.$$

5. **(C)** From Section 8.4, solution is
   $$\dfrac{1}{e^{-4x^2}}\int \left[e^{-4x^2}x\,dx + C\right] = -\dfrac{1}{8} + Ce^{4x^2}.$$

   > Letting $u = -4x^2$ and $du = -8x\,dx$, we have $\displaystyle\int e^{-4x^2}x\,dx = -\dfrac{1}{8}\int e^u\,du$
   > $$= -\dfrac{1}{8}e^{-4x^2} + C.$$

6. **(D)** The characteristic equation is
   $$D^2 - 4D + 3 = 0 \Rightarrow (D - 3)(D - 1) = 0.$$

7. **(A)** $y = \dfrac{x^3}{3} + C$. $4 = \dfrac{27}{3} + C \Rightarrow C = -5$.

8. **(B)** Separate the variables. $e^y\,dy = dx \Rightarrow e^y = x + C \Rightarrow y = \ln|x + C|$.

9. **(D)** $\dfrac{dy}{dx} = y$ is describing a function that is its own derivative.

    $ax + b$. ($y_p = a$ would not be adequate since left-hand side would be 0.) $y_p = ax + b \Rightarrow a = -2$ and $b = 0$. Hence solution is $y = c_1 + c_2 e^{3x} - 2x$.

10. **(C)** Recall that $\dfrac{d}{dx}(\tan^{-1} x) = \dfrac{1}{1 + x^2}$. Hence $y = \tan^{-1} x + C.$ $2 = \tan^{-1} 1 + C \Rightarrow$ $C = 2 - \dfrac{\pi}{4}.$

    > Note that choice (B) is a particular solution, not a general solution.

11. **(E)** Homogeneous solution is $y_c = c_1 + c_2 e^{3x}$. To find y_p, assume y_p has the form

12. **(B)** By inspection, a particular solution is $y = 1$ ($y'' = y' = 0$).

SUPPLEMENTAL PROBLEMS

This section contains a series of calculus or calculus-related problems in no particular order of topics. The comments provided with the solutions will direct you to the section of this book that relates to the material in the question. Those questions marked (*) involve topics usually found in an advanced calculus course or on the Calculus BC Advanced Placement Examination.

1. Given the two curves $5y - 2x + y^3 - x^2y = 0$ and $5x + 2y + x^4 - x^3y^2 = 0$.
   (a) Show that both curves pass through the origin.
   (b) Prove that the tangents to the curves at the origin are perpendicular.

2. Sketch the graph of $f(x) = 2\sin x + \cos x$ over the interval $[0, 2\pi]$. Then
   (a) Identify all maximum and minimum points.
   (b) Identify all x-intercepts.
   (c) Find all points of inflection.
   (d) Identify all asymptotes.
   (e) Identify the intervals where the function is (i) concave upward; (ii) concave downward.
   (f) Identify the intervals where the function is (i) increasing; (ii) decreasing.

3. Find the area of the region bounded by the graph of $y = 4 - x^2$ and $y = 0$.

4. A bacterial population is growing at a rate equal to 10% of its population each day. If its initial size was 10,000 organisms, how many bacteria were present after 10 days?

5. The radius of a sphere is found to be 4.01 inches rather than 4 inches. Use differentials to find
   (a) the approximate error in the surface area
   (b) the approximate error in the volume

6. A particle starts from rest at a point 13 units to the left of the origin. Its acceleration at any time $t \geq 0$ is $a(t) = 2t + 4$ ft./sec.^{2}. Find its velocity and position functions.

7. Let $f(x) = x^2 + 2x - 7$ on the interval $[-2, 3]$. Find the point(s) on the curve where the tangent line is parallel to the segment joining the point where $x = -2$ to the point where $x = 3$.

8. In each case, determine whether or not the given function is continuous at the indicated point.
   (a) $f(x) = \begin{cases} \dfrac{x^2 - 16}{x + 4} & \text{if } x \neq -4 \\ -8 & \text{if } x = -4 \end{cases}$
   Continuous at $x = -4$?
   (b) $g(x) = \begin{cases} x^3 - 2x + 5 & \text{if } x \geq 1 \\ 2x^2 + 3 & \text{if } x < 1 \end{cases}$
   Continuous at $x = 1$?
   (c) $f(x) = \begin{cases} 5\sin x & \text{if } x > \pi/4 \\ 5\cos x & \text{if } x \leq \pi/4 \end{cases}$
   Continuous at $x = \pi/4$?
   (d) $w(x) = \begin{cases} 1 & \text{if } x \text{ is rational} \\ 0 & \text{if } x \text{ is irrational} \end{cases}$
   Continuous at any point?

9. Find the derivative for each function.
   (a) $f(x) = e^{\cos x}$
   (b) $g(x) = \ln\left(\dfrac{x^2 + 4}{x^2 + 1}\right)^3$
   (c) $d(x) = \sin^2(4x - 2)$
   (d) $t(x) = 19^x$

(e) $s(x) = \sin^{-1} 4x$

(f) $r(x) = (\ln x^5)(\tan 3x)$

10. Evaluate the integrals.

(a) $\int \dfrac{x^5 + 1}{x^3}\,dx$

(b) $\int \dfrac{x^6}{x^7 + 5}\,dx$

(c) $\int \cos 4x\,dx$

(d) $\int_0^2 xe^x\,dx$

(e) $\int_0^{\pi/2} \sin x\sqrt{1 + \cos x}\,dx$

(f) $\int_0^{10} \dfrac{dx}{\sqrt{100 - x^2}}$

11. Given a rectangular sheet of cardboard, 16 in. by 10 in., you are asked to cut off identical squares from the four corners of the sheet, and then bend up the sides to form an open rectangular box. What is the maximum volume of such a box?

12. Find the *mean value* for each function in the designated interval:

(a) $f(x) = \sin x$ on $[0, \pi/4]$

(b) $g(x) = x^3 + x^{-1}$ on $[1, e^4]$

13. Find the area of the region bounded by the parabola $y^2 = x$ and the line $y = x - 2$.

14. Evaluate:

(a) $\lim\limits_{x\to 9} \dfrac{x^2 - 81}{x - 9}$

(b) $\lim\limits_{h\to 0} \dfrac{(x + h)^{1/3} - x^{1/3}}{h}$

(c) $\lim\limits_{x\to 0} \dfrac{1 - e^{20x}}{4x}$

(d) $\lim\limits_{x\to 0} \dfrac{x - \tan x}{x - \sin x}$

15. Find the volume of the ellipsoid of revolution obtained by revolving the ellipse $\dfrac{x^2}{4} + \dfrac{y^2}{25} = 1$

(a) about the x-axis

(b) about the y-axis

16. Water is pouring into a cylindrical bowl of height 6 feet and radius 4 feet at a rate of 5 cubic feet per minute.

(a) At what rate does the surface of the water rise?

(b) How long does it take to fill the bowl?

17. A particle moves on the x-axis so that its velocity at any time $t \geq 0$ is given by $v(t) = 4\cos 2t$. At $t = 0$, the particle is 5 units to the right of the origin.

(a) Find all values of $t \in [0, \pi]$ for which the particle is moving to the left.

(b) What is the acceleration of the particle at $t = \pi/12$?

(c) What is the location of the particle at $t = \pi/8$?

(d) What is the average value for the position function on the interval $[0, \pi/2]$?

18. Consider the functions $f(x) = x^4 + 1$ and $g(x) = \ln 4x$, both defined on $[1, \infty)$. Calculate

(a) $(f \circ g)(x)$

(b) $(g \circ f)(x)$

(c) $(f \circ g)'(x)$

(d) $(g \circ f)'(x)$

(e) $f^{-1}(x)$

(f) $g^{-1}(x)$

(g) $(f^{-1})'(x)$

(h) $(g^{-1})'(x)$

(i) The slope of the tangent line to $(g^{-1} \circ f^{-1})(x)$ at the point where $x = 2$.

19. Evaluate

(a) $\int_0^2 f(x)\,dx$, given that

$$f(x) = \begin{cases} 4 + \sin x, & x \geq 1 \\ \frac{3}{2}x + \frac{5}{2}, & x < 1 \end{cases}$$

(b) $\int_0^1 \sqrt{x^2 - 10x + 25}\,dx$

20. Consider the function $t(x) = \sin x \cos x$ on the interval $[0, \pi]$.

(a) Sketch a graph of the function.

(b) Find the maximum and minimum values of the function.

(c) Find the coordinates of all points of inflection.

(d) On the interval $[0, \pi/2]$, find the area of the region bordered by the graph of the function and the line $y = 0$.

(e) What is the period of this function?

21. Given $4x^2 + 2e^{2x} - 6 = \int_k^x f(t)\,dt$, where k is a constant.
    (a) Find $f(x)$.
    (b) Show that k must be a solution of the equation $4x^2 + 2e^{2x} = 6$.

22. The region bounded by the graphs of $xy = 1$, $x = 1$, $x = e$, and $y = 0$ is revolved about the line $y = -1$. Find the volume of the solid generated.

23. The region bounded by $y = \dfrac{1}{x}\sin x$, $y = 0$, and $x = \dfrac{\pi}{2}$ is revolved about the y-axis. Use the shell method to find the volume of the solid generated. $\Big[$ Note that $y = \dfrac{1}{x}\sin x$ is continuous at $x = 0$ if its value is defined to be $\lim_{x \to 0}\left(\dfrac{1}{x}\sin x\right) = 1.\Big]$

24. Find the equation of the tangent line to $y = \tan\left(\dfrac{\pi x y^2}{16}\right)$ at $(1, 2)$.

*25. Evaluate
    (a) $\displaystyle\int_2^\infty \dfrac{dx}{x^2}$
    (b) $\displaystyle\int_7^{15} \dfrac{dx}{(x-7)^{2/3}}$
    (c) $\displaystyle\int_0^2 \dfrac{2x\,dx}{(x^2-1)^2}$

*26. A particle moves in a plane according to

$x = e^{5\sin t}$, $y = e^{5\cos t}$. Find the speed of the particle at time $t = 0$.

*27. Find the area of the first quadrant region of the cardioid $r = 1 - \cos\theta$.

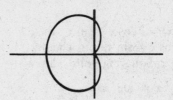

*28. Find the arc length of the curve $y = x^{3/2}$ from $x = 0$ to $x = 4$.

*29. Find the surface area of the solid of revolution generated by revolving $y = \sqrt{1 - x^2}$, $\frac{1}{4} \le x \le \frac{3}{4}$, about the x-axis.

*30. Determine the radius of convergence for
    (a) $\displaystyle\sum_{n=0}^\infty nx^n$
    (b) $\displaystyle\sum_{n=2}^\infty \dfrac{x^n}{(n-2)!}$

*31. Write the first four terms in the Maclaurin Series for $(1 + x)^{1/2}$.

*32. (a) Find the general solution of the differential equation $(D^2 + 1)y = 2x$.
    (b) Find the particular solution of the equation if the graph passes through $(0, 3)$ with a slope equal to 2.

*33. Solve the differential equation
    $x(1 + y)\,dx + (1 + x^2)\,dy = 0$.

Solutions and Comments for Supplemental Problems

1. (a) $(0, 0)$ satisfies both equations.
   (b) Differentiating the first equation implicitly,
   $$5\dfrac{dy}{dx} - 2 + 3y^2 \times \dfrac{dy}{dx} - x^2 \times \dfrac{dy}{dx} - 2xy = 0$$
   $$\dfrac{dy}{dx} = \dfrac{2xy + 2}{3y^2 - x^2 + 5}.$$ At $(0, 0)$, slope of tangent line is $2/5$. Differentiating the second equation implicitly yields

   $$\dfrac{dy}{dx} = \dfrac{3x^2y^2 - 4x^3 - 5}{2 - 2x^3y}.$$ At $(0, 0)$, slope of tangent line is $-5/2$.

   > Tangent lines (Section 4.1); implicit differentiation (Section 3.5). Two lines are perpendicular if the product of their slopes is -1.

2.

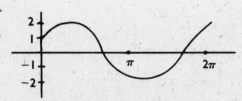

(a) $f'(x) = 2\cos x - \sin x$; this is zero when $2\cos x = \sin x$, or when $2\sqrt{1 - \sin^2 x} = \sin x$. Squaring both sides, $4(1 - \sin^2 x) = \sin^2 x \Rightarrow \sin^2 x = \dfrac{4}{5} \Rightarrow x = \sin^{-1}\dfrac{2\sqrt{5}}{5}$ or $\pi + \sin^{-1}\dfrac{2\sqrt{5}}{5}$. Max. at $x = \sin^{-1}\dfrac{2\sqrt{5}}{5}$; min. at $x = \pi + \sin^{-1}\dfrac{2\sqrt{5}}{5}$.

(b) x-intercepts occur when $2\sin x = -\cos x$ $\left(\text{i.e., when } x = \pi - \sin^{-1}\dfrac{\sqrt{5}}{5} \text{ and } x = 2\pi - \sin^{-1}\dfrac{\sqrt{5}}{5}\right)$.

(c) $f''(x) = -2\sin x - \cos x$. Inflection points are same as x-intercepts.

(d) No asymptotes.

(e) $f''(x) \le 0$ when $x \in \left[0, \pi - \sin^{-1}\dfrac{\sqrt{5}}{5}\right]$ or when $x \in \left[2\pi - \sin^{-1}\dfrac{\sqrt{5}}{5}, 2\pi\right]$. $f''(x) \ge 0$ when $x \in \left[\pi - \sin^{-1}\dfrac{\sqrt{5}}{5}, 2\pi - \sin^{-1}\dfrac{\sqrt{5}}{5}\right]$.

> Concave downward when $f''(x) \le 0$; concave upward when $f''(x) \ge 0$.

(f) $f'(x) \ge 0$ when $x \in \left[0, \sin^{-1}\dfrac{2\sqrt{5}}{5}\right]$ or when $x \in \left[\pi + \sin^{-1}\dfrac{2\sqrt{5}}{5}, 2\pi\right]$. $f'(x) \le 0$ when $x \in \left[\sin^{-1}\dfrac{2\sqrt{5}}{5}, \pi + \sin^{-1}\dfrac{2\sqrt{5}}{5}\right]$.

> f increasing when $f'(x) \ge 0$. f decreasing when $f'(x) \le 0$.

> Curve sketching (Section 4.2). This section includes information about testing for max., min., points of inflection, etc.

3. $\displaystyle\int_{-2}^{2}(4 - x^2)\,dx = \left(4x - \dfrac{x^3}{3}\right)\Big|_{-2}^{2} = \dfrac{32}{3}$ (sq. units).

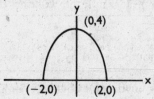

> Section 6.5.

4. If p represents the amount present at any time t, then $\dfrac{dp}{dt} = 0.1p \Rightarrow p(t) = Ce^{0.1t}$. $p(0) = 10000 \Rightarrow C = 10000$. Hence $p(t) = 10000e^{0.1t}$; $p(10) = 10000e^1 = 27183$.

> Section 5.3.

5. (a) $S = 4\pi r^2 \Rightarrow dS = 8\pi r\,dr$. If $r = 4$ and $dr = 0.01$, then $dS = 8\pi(4)(0.01)$, or about 1.005 sq. in.

(b) $V = \frac{4}{3}\pi r^3 \Rightarrow dV = 4\pi r^2\,dr$. If $r = 4$ and $dr = 0.01$, then $dV = 4\pi(16)(.01)$, or about 2.01 cu. in.

> Section 4.1.

6. $v(t) = t^2 + 4t + v_0$; $v(0) = 0$, $v_0 = 0$; hence $v(t) = t^2 + 4t$.

$s(t) = \dfrac{t^3}{3} + 2t^2 + s_0$; $s(0) = -13 = s_0$;

hence $s(t) = \dfrac{t^3}{3} + 2t^2 - 13$.

> Section 5.2.

7. $f'(x) = 2x + 2; \dfrac{f(3) - f(-2)}{3 - (-2)} = 2c + 2;$

$\dfrac{8 - (-7)}{5} = 3 = 2c + 2 \Rightarrow c = \dfrac{1}{2}.$

The only point with the desired property is $\left(\dfrac{1}{2}, -5\dfrac{3}{4}\right).$

Mean Value Theorem (Section 3.9).

---

8. (a) Yes; (b) No; (c) Yes; (d) No.

Section 2.5. In (d), we have a function that is defined everywhere, but which is not continuous anywhere.

---

9. (a) $f'(x) = (-\sin x)e^{\cos x}.$

(b) $g'(x) = 3\left[\dfrac{2x}{x^2 + 4} - \dfrac{2x}{x^2 + 1}\right].$

(c) $d'(x) = 2[\sin(4x - 2)\cos(4x - 2)]4$
$= 8\sin(4x - 2)\cos(4x - 2).$

(d) If $y = 19^x$, then $\ln y = x(\ln 19) \Rightarrow$
$\dfrac{1}{y} \times \dfrac{dy}{dx} = \ln 19 \Rightarrow \dfrac{dy}{dx} = t'(x) = 19^x \ln 19.$

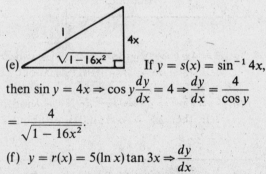

(e) If $y = s(x) = \sin^{-1} 4x$,

then $\sin y = 4x \Rightarrow \cos y \dfrac{dy}{dx} = 4 \Rightarrow \dfrac{dy}{dx} = \dfrac{4}{\cos y}$

$= \dfrac{4}{\sqrt{1 - 16x^2}}.$

(f) $y = r(x) = 5(\ln x)\tan 3x \Rightarrow \dfrac{dy}{dx}$

$= 5\left[(\ln x)3\sec^2 3x + (\tan 3x)\dfrac{1}{x}\right].$

(a) Section 3.4;
(b) $g(x) = 3[\ln(x^2 + 4) - \ln(x^2 + 1)]$
(Section 3.7);
(c) Let $y = u^2$, $u = \sin w$, $w = 4x - 2$
(Section 3.4);
(d) Section 3.7;
(e) Section 3.6;
(f) Section 3.3.

---

10. (a) $\dfrac{x^3}{3} - \dfrac{1}{2x^2} + C;$

(b) $\dfrac{1}{7}\ln|x^7 + 5| + C;$

(c) $\dfrac{1}{4}\sin 4x + C;$

(d) $(xe^x - e^x)\Big|_0^2 = (2e^2 - e^2) - (0 - 1)$
$= e^2 + 1;$

(e) If $u = 1 + \cos x$, then $du = -\sin x\, dx.$

$-\dfrac{2}{3}(1 + \cos x)^{3/2}\Big|_0^{\pi/2} = -\dfrac{2}{3}(1 - 2^{3/2})$

$= -\dfrac{2}{3}(1 - 2\sqrt{2});$

(f) $\sin^{-1}\left(\dfrac{x}{10}\right)\Big|_0^{10} = \sin^{-1}(1) - \sin^{-1}(0) = \dfrac{\pi}{2}$

(a) Write as $\displaystyle\int (x^2 + x^{-3})\, dx$ (Section 5.1).

(b) If $u = x^7 + 5$, integral form as $\dfrac{1}{7}\displaystyle\int \dfrac{du}{u}.$

(c) Section 5.4.
(d) Integration by parts (Section 5.5).
(e) With indicated substitution, integral form is $\displaystyle\int -u^{1/2}\, du.$

(f) Section 5.1, formula (18).

---

11. If $x = $ length of side of cut-off square, then
$V(x) = x(16 - 2x)(10 - 2x)$
$\quad = 4x^3 - 52x^2 + 160x.$
$V'(x) = 4(3x^2 - 26x + 40)$
$\quad = 4(3x - 20)(x - 2) = 0$
when $x = 20/3$ or $x = 2$. Max. occurs when
$x = 2; V(2) = 144$ (cu. in.).

Sections 4.2, 4.3.

---

12. (a) $\dfrac{\displaystyle\int_0^{\pi/4} \sin x\, dx}{\dfrac{\pi}{4} - 0} = \dfrac{4}{\pi}(-\cos x)\Big|_0^{\pi/4} = \dfrac{8 - 4\sqrt{2}}{2\pi}.$

(b) $\dfrac{\displaystyle\int_1^{e^4} (x^3 + x^{-1})\, dx}{e^4 - 1} = \dfrac{\left[\dfrac{x^4}{4} + \ln x\right]_1^{e^4}}{e^4 - 1}$

$= \dfrac{e^{16} + 15}{4(e^4 - 1)}.$

Mean Value Theorem for Integrals (Section 6.3).

13.

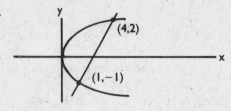

$$\int_{-1}^{2} \left[(y + 2) - y^2 \right] dy = \left(\frac{y^2}{2} + 2y - \frac{y^3}{3} \right) \Big|_{-1}^{2}$$

$$= \frac{9}{2} \text{ (sq. units)}.$$

(Section 6.5). If one integrates with respect to x, one must calculate

$$\int_{0}^{1} \left[\sqrt{x} - (-\sqrt{x}) \right] dx$$

$$+ \int_{1}^{4} \left[\sqrt{x} - (x - 2) \right] dx.$$

14. (a) $\lim_{x \to 9} (x + 9) = 18$;

(b) $\frac{1}{3} x^{-2/3}$;

(c) $\lim_{x \to 0} \left(\frac{-20 e^{20x}}{4} \right) = \frac{-20}{4} = -5$;

(d) $\lim_{x \to 0} \frac{1 - \sec^2 x}{1 - \cos x} = \lim_{x \to 0} \frac{(\cos x + 1)(\cos x - 1)}{\cos^2 x (1 - \cos x)}$

$$= -2.$$

(a) Section 2.1.
(b) This is the definition of the derivative of $y = x^{1/3}$ (Section 3.1).
(c) L'Hopital's Rule (Section 3.11).
(d) L'Hopital's Rule still leaves indeterminate form. Note however that

$$\frac{1 - \sec^2 x}{1 - \cos x} = \frac{\dfrac{\cos^2 x - 1}{\cos^2 x}}{1 - \cos x}$$

$$= \frac{(\cos x + 1)(\cos x - 1)}{\cos^2 x (1 - \cos x)}.$$

15.

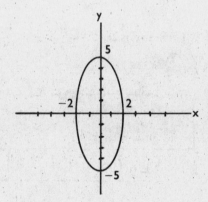

(a) $2 \int_{0}^{2} \pi 25 \left(1 - \frac{x^2}{4} \right) dx = 2\pi \left(25x - \frac{25x^3}{12} \right) \Big|_{0}^{2}$

$$= \frac{200\pi}{3} \text{ (cu. units)}.$$

(b) $2 \int_{0}^{5} \pi 4 \left(1 - \frac{y^2}{25} \right) dy = 2\pi \left(4y - \frac{4y^3}{75} \right) \Big|_{0}^{5}$

$$= \frac{80\pi}{3} \text{ (cu. units)}.$$

Section 6.6.

16.

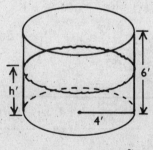

$V = \pi r^2 h$

(a) $V = 16\pi h \Rightarrow \dfrac{dV}{dt} = 16\pi \dfrac{dh}{dt} \Rightarrow \dfrac{dh}{dt} = \dfrac{\dfrac{dv}{dt}}{16\pi}$

$$= \frac{5}{16\pi} \text{ (ft./min.)}.$$

(b) Volume of bowl $= 4^2 \pi \times 6 = 96\pi$ (cu. ft.).

$$\frac{96\pi}{5} = 60.3 \text{ (minutes)}.$$

Section 4.5.

17. (a) $v(t) = 4 \cos 2t$ in $[0, \pi]$. Particle moving to

left when $\cos 2t < 0 \Rightarrow \dfrac{\pi}{2} < 2t < \dfrac{3\pi}{2} \Rightarrow \dfrac{\pi}{4} <$

$t < \dfrac{3\pi}{4}$.

(b) $a(t) = -8\sin 2t$; $a(\pi/12) = -4$.

(c) $s(t) = 2\sin 2t + 5$; $s(\pi/8) = \sqrt{2} + 5$.

(d) $\dfrac{\displaystyle\int_0^{\pi/2} (2\sin 2t + 5)\,dt}{\dfrac{\pi}{2} - 0} = \dfrac{2}{\pi}(-\cos 2t + 5t)\Big|_0^{\pi/2}$

$= \dfrac{4}{\pi} + 5.$

Sections 4.4, 5.2, 6.3.

---

18. (a) $(f \circ g)(x) = (\ln 4x)^4 + 1$;

(b) $(g \circ f)(x) = \ln(4x^4 + 4)$;

(c) $(f \circ g)'(x) = \dfrac{4}{x}(\ln 4x)^3$;

(d) $(g \circ f)'(x) = \dfrac{16x^3}{4x^4 + 4}$;

(e) $f^{-1}(x) = (x-1)^{1/4}$, $x \in [2, \infty)$;

(f) $g^{-1}(x) = \frac{1}{4}e^x$, $x \in [\ln 4, \infty)$;

(g) $(f^{-1})'(x) = \frac{1}{4}(x-1)^{-3/4}$;

(h) $(g^{-1})'(x) = \frac{1}{4}e^x$;

(i) $(g^{-1} \circ f^{-1})(x) = \frac{1}{4}e^{(x-1)^{\frac{1}{4}}} \Rightarrow$
$(g^{-1} \circ f^{-1})'(x) = \frac{1}{4}e^{(x-1)^{\frac{1}{4}}}[\frac{1}{4}(x-1)^{-3/4}] \Rightarrow$
$(g^{-1} \circ f^{-1})'(2) = \dfrac{e}{16}.$

Sections 1.2, 1.4, 3.4, 3.6.

---

19. (a) $\displaystyle\int_1^2 (4 + \sin \pi x)\,dx + \int_0^1 \left(\dfrac{3}{2}x + \dfrac{5}{2}\right)dx$

$= \left(4x - \dfrac{1}{\pi}(\cos \pi x)\right)\Big|_1^2 + \left[\dfrac{3x^2}{4} + \dfrac{5}{2}x\right]_0^1$

$= \dfrac{29}{4} - \dfrac{2}{\pi}.$

(b) $\displaystyle\int_0^1 \sqrt{(x-5)^2} = \int_0^1 (5-x)\,dx$

$= \left(5x - \dfrac{x^2}{2}\right)\Big|_0^1 = 4\frac{1}{2}.$

Section 6.3. In (b), note that $x - 5 < 0$ in $[0, 1]$. Hence $\sqrt{(x-5)^2} = 5 - x$.

---

20. (a)

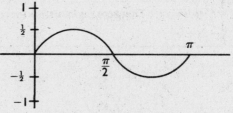

(b) $t(x) = \sin x \cos x = \frac{1}{2}\sin 2x$.

$t'(x) = \cos 2x = 0$ when $x = \dfrac{\pi}{4}$ and $x = \dfrac{3\pi}{4}$.

Max. $= t\left(\dfrac{\pi}{4}\right) = \dfrac{1}{2}.$

Min. $= t\left(\dfrac{3\pi}{4}\right) = -\dfrac{1}{2}.$

(c) $\left(\dfrac{\pi}{2}, 0\right)$ is a point of inflection.

(d) $\dfrac{1}{2}\displaystyle\int_0^{\pi/2} \sin 2x\,dx = \dfrac{1}{2}\left[\dfrac{-\cos 2x}{2}\right]\Big|_0^{\pi/2} = \dfrac{1}{2}.$

(e) Period $= \pi$.

Section 4.2, 6.1, and 1.6. The trigonometric identity $\sin 2x = 2\sin x \cos x$ is very useful in this problem.

---

21. (a) $f(x) = 8x + 4e^{2x}$;

(b) Letting $x = k$, we have $4k^2 + 2e^{2k} - 6 =$

$\displaystyle\int_k^k f(t)\,dt = 0 \Rightarrow 4k^2 + 2e^{2k} = 6.$

Section 6.4. $\dfrac{d}{dx}\displaystyle\int_k^x f(t)\,dt = f(x)$.

22.

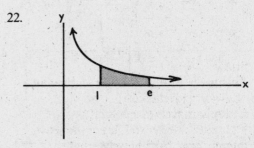

$\pi \displaystyle\int_1^e \left[\left(1 + \dfrac{1}{x}\right)^2 - 1^2\right]dx = \pi \int_1^e \left(\dfrac{2}{x} + \dfrac{1}{x^2}\right)dx$

$= \pi \left[2\ln x - \dfrac{1}{x}\right]\Big|_1^e$

$= \pi \left(3 - \dfrac{1}{e}\right).$

Section 6.6.

23. $\int_0^{\pi/2} 2\pi x \times \frac{1}{x}\sin x \, dx = \int_0^{\pi/2} 2\pi \sin x \, dx$

$$= -2\pi \cos x \Big|_0^{\pi/2}$$

$$= 0 + 2\pi = 2\pi.$$

Section 6.7.

24. $\dfrac{dy}{dx} = \left(\sec^2\left(\dfrac{\pi x y^2}{16}\right)\right)\dfrac{\pi}{16}\left[x 2y \dfrac{dy}{dx} + y^2\right].$

At $(1,2)$, $\dfrac{dy}{dx} = \dfrac{\pi}{2}\left[\dfrac{dy}{dx} + 1\right] \Rightarrow \dfrac{dy}{dx} = \dfrac{\pi}{2-\pi}.$

Equation of tangent line at $(1,2)$ is $y - 2 = $

$\dfrac{\pi}{2-\pi}(x - 1).$

Section 3.5.

25. (a) $\displaystyle\lim_{b\to\infty}\left[-\dfrac{1}{x}\Big|_2^b\right] = \lim_{b\to\infty}\left(-\dfrac{1}{b} + \dfrac{1}{2}\right) = \dfrac{1}{2}.$

(b) $\displaystyle\lim_{a\to7^+}\int_a^{15}\dfrac{dx}{(x-7)^{2/3}} = \lim_{a\to7^+} 3(x-7)^{1/3}\Big|_a^{15}$

$= \displaystyle\lim_{a\to7^+}\left[3\times8^{1/3} - 3(a-7)^{1/3}\right] = 6.$

(c) $\displaystyle\lim_{b\to1^-}\int_0^b\dfrac{2x}{(x^2-1)^2}dx + \lim_{a\to1^+}\int_a^2\dfrac{2x}{(x^2-1)^2}dx$

$= \displaystyle\lim_{b\to1^-}\left[\dfrac{-1}{x^2-1}\right]_0^b + \lim_{a\to1^+}\left[\dfrac{-1}{x^2-1}\right]_a^0$

$= \displaystyle\lim_{b\to1^-}\left[\dfrac{-1}{b^2-1} - 1\right] + \lim_{a\to1^+}\left[1 - \dfrac{-1}{a^2-1}\right].$

Neither limit exists; integral is divergent.

Section 6.10.

26. $x' = 5(\cos t)e^{5\sin t}$, $y' = -5(\sin t)e^{5\cos t}$.
$x'(0) = 5$; $y'(0) = 0 \Rightarrow$ speed $= 5$.

Section 4.4A.

27. $\displaystyle\int_0^{\pi/2}\dfrac{1}{2}r^2\,d\theta = \dfrac{1}{2}\int_0^{\pi/2}(1-\cos\theta)^2\,d\theta$

$$= \dfrac{1}{2}\int_0^{\pi/2}(1 - 2\cos\theta + \cos^2\theta)\,d\theta$$

$$= \dfrac{1}{2}\int_0^{\pi/2}\left(\dfrac{3}{2} - 2\cos\theta + \dfrac{1}{2}\cos 2\theta\right)d\theta$$

$$= \dfrac{1}{2}\left[\dfrac{3}{2}\theta - 2\sin\theta + \dfrac{1}{4}\sin 2\theta\right]_0^{\pi/2}$$

$$= \dfrac{3\pi}{8} - 1.$$

> Section 6.5A. The trigonometric identity $\cos^2\theta = \frac{1}{2}(1 + \cos 2\theta)$ is helpful in this problem.

28. $\dfrac{dy}{dx} = \dfrac{3}{2}x^{1/2}$. The desired length is

$\displaystyle\int_0^4\sqrt{1 + \dfrac{9}{4}x}\,dx = \dfrac{4}{9}\int_0^4\sqrt{1 + \dfrac{9}{4}x}\left(\dfrac{9}{4}dx\right)$

$$= \dfrac{8}{27}\left(1 + \dfrac{9}{4}x\right)^{3/2}\Big|_0^4$$

$$= \dfrac{8}{27}(10^{3/2} - 1).$$

Section 6.8.

29. $\dfrac{dy}{dx} = \dfrac{-x}{\sqrt{1-x^2}}.$

$\displaystyle\int_{1/4}^{3/4} 2\pi\sqrt{1-x^2}\sqrt{1 + \dfrac{x^2}{1-x^2}}\,dx = \int_{1/4}^{3/4} 2\pi\,dx$

$$= 2\pi x\Big|_{1/4}^{3/4}$$

$$= 2\pi\left(\dfrac{3}{4} - \dfrac{1}{4}\right)$$

$$= \pi.$$

Section 6.9.

30. (a) $\displaystyle\lim_{n\to\infty}\left|\dfrac{a_{n+1}}{a_n}\right| = \lim_{n\to\infty}\left(\dfrac{n+1}{n}\right)|x| = |x|.$

Hence, radius of convergence is 1.

(b) $\displaystyle\lim_{n\to\infty}\left|\dfrac{a_{n+1}}{a_n}\right| = \lim_{n\to\infty}\left(\dfrac{1}{n-1}\right)|x| = 0.$

Hence, series converges for all x.

$$\boxed{\text{Section 7.3.}}$$

31. $f(x) = (1 + x)^{1/2} \Rightarrow f(0) = 1$.
$f'(x) = \frac{1}{2}(1 + x)^{-1/2} \Rightarrow f'(0) = \frac{1}{2}$.
$f''(x) = -\frac{1}{4}(1 + x)^{-3/2} \Rightarrow f''(0) = -\frac{1}{4}$.
$f'''(x) = \frac{3}{8}(1 + x)^{-5/2} \Rightarrow f'''(0) = \frac{3}{8}$. Hence,
$(1 + x)^{1/2} = 1 + \frac{1}{2}x - \frac{1}{8}x^2 + \frac{1}{16}x^3 + \cdots$.

$$\boxed{\text{Section 7.3.}}$$

32. (a) The characteristic $D^2 + 1 = 0$ has solutions $0 + i$ and $0 - i$. Hence the solution to $(D^2 + 1)y = 0$ is $y = C_1 \cos x + C_2 \sin x$. Now assuming $y_p = ax + b$, we have $y_p' = a$ and $y_p'' = 0$. Then $ax + b = 2x \Rightarrow a = 2$ and $b = 0$. Hence the general solution is $y = C_1 \cos x + C_2 \sin x + 2x$.

(b) If graph contains $(0, 3)$, then $C_1 = 3$. Also, $y' = -C_1 \sin x + C_2 \cos x + 2$. Since $y' = 2$ when $x = 0$, we have $C_2 = 0$. Particular solution is $y = 3 \cos x + 2x$.

$$\boxed{\text{Sections 8.5, 8.6.}}$$

33. $\dfrac{x \, dx}{1 + x^2} = \dfrac{-dy}{1 + y} \Rightarrow \displaystyle\int \dfrac{x \, dx}{1 + x^2} = -\int \dfrac{dy}{1 + y}$

$\Rightarrow \dfrac{1}{2} \ln(1 + x^2) = -\ln|1 + y| + C$

$\Rightarrow \dfrac{1}{2} \ln(1 + x^2) + \ln|1 + y| = C$

$\Rightarrow \ln \sqrt{1 + x^2} + \ln|1 + y| = C$

$\Rightarrow \ln\left[\sqrt{1 + x^2}|1 + y|\right] = C$

$\Rightarrow |1 + y|\sqrt{1 + x^2} = e^C$

$\Rightarrow |1 + y|\sqrt{1 + x^2} = C_1$.

$$\boxed{\text{Section 8.1.}}$$

CALCULUS FORMULAS AND THEOREMS FOR REFERENCE

I. General Formulas and Theorems

1. $f'(x) = \lim\limits_{h \to 0} \dfrac{f(x+h) - f(x)}{h}$

2. $(fg)' = fg' + gf'$

3. $\left(\dfrac{f}{g}\right)' = \dfrac{gf' - fg'}{g^2}$

4. $\dfrac{d}{dx} f(g(x)) = f'(g(x))g'(x)$

5. *Mean Value Theorem:* If f is continuous on $[a, b]$ and differentiable on (a, b), then there is at least one number $c \in (a, b)$ such that

$$\dfrac{f(b) - f(a)}{b - a} = f'(c)$$

6. *L'Hopital's Rule:* If $\lim\limits_{x \to a} \dfrac{f(x)}{g(x)}$ is indeterminate of the form $\dfrac{0}{0}$ or $\dfrac{\infty}{\infty}$, and if $\lim\limits_{x \to a} \dfrac{f'(x)}{g'(x)}$ exists, then

$$\lim_{x \to a} \dfrac{f(x)}{g(x)} = \lim_{x \to a} \dfrac{f'(x)}{g'(x)}.$$

7. $f'(a) = 0$ and $f''(a) > 0 \Rightarrow f$ has a relative min. at $x = a$. $f'(a) = 0$ and $f''(a) < 0 \Rightarrow f$ has a relative max. at $x = a$.

8. $\displaystyle\int kf(x)\,dx = k \int f(x)\,dx$

9. $\displaystyle\int u\,dv = uv - \int v\,du$

10. $\displaystyle\int f(x)\,dx = F(x) + C$, where $F'(x) = f(x)$

11. $\displaystyle\int_a^b f(x)\,dx = \lim_{n \to \infty} \left(\dfrac{b - a}{n}\right) \sum_{k=0}^{n-1} f\left(a + \dfrac{k}{n}(b - a)\right)$

12. $\displaystyle\int_b^a f(x)\,dx = - \int_a^b f(x)\,dx$

13. *Mean Value Theorem for Integrals:* There exists at least one $c \in [a, b]$ such that

$$f(c) = \dfrac{\displaystyle\int_a^b f(x)\,dx}{b - a}.$$

$f(c)$ is the *mean value* of f with respect to x on $[a, b]$.

14. $\displaystyle\int_a^b f(x)\,dx = F(b) - F(a)$, where $F'(x) = f(x)$.

15. $\dfrac{d}{dx} \displaystyle\int_a^x f(x)\,dx = f(x)$

16. If f and g are continuous such that $f(x) \geq g(x)$ on $[a, b]$, then the area between the curves is

$$\int_a^b [f(x) - g(x)]\,dx.$$

17. *Volumes:*

Disk method about x-axis:

$$\int_a^b \pi y^2 \, dx$$

Shell method about y-axis:

$$\int_a^b 2\pi xy \, dx$$

18. *Arc Length:*

$$\int_a^b \sqrt{1 + \left(\frac{dy}{dx}\right)^2} \, dx$$

Surface Area
About x-axis:

$$\int_a^b 2\pi f(x)\sqrt{1 + (f'(x))^2} \, dx$$

II. Specific Differentiation Formulas

1. $(x^n)' = nx^{n-1}$

2. $(\sin x)' = \cos x$

3. $(\cos x)' = -\sin x$

4. $(\tan x)' = \sec^2 x$

5. $(\csc x)' = -\csc x \cot x$

6. $(\sec x)' = \sec x \tan x$

7. $(\cot x)' = -\csc^2 x$

8. $(\ln x)' = \dfrac{1}{x}$

9. $(\log_b x)' = \dfrac{1}{x}\log_b e$

10. $(b^x)' = b^x \ln b$

11. $(e^x)' = e^x$

12. $(\sin^{-1} x)' = \dfrac{1}{\sqrt{1 - x^2}}$

13. $(\cos^{-1} x)' = -\dfrac{1}{\sqrt{1 - x^2}}$

14. $(\tan^{-1} x)' = \dfrac{1}{1 + x^2}$

III. Table of Integrals

(The constant of integration C is omitted.)

1. $\displaystyle\int u^n \, du = \frac{u^{n+1}}{n+1}, \ n \neq -1$

2. $\displaystyle\int \frac{du}{u} = \log |u|$

3. $\displaystyle\int e^u \, du = e^u$

4. $\displaystyle\int \sin u \, du = -\cos u$

5. $\displaystyle\int \cos u \, du = \sin u$

6. $\displaystyle\int \sec^2 u \, du = \tan u$

7. $\displaystyle\int \csc^2 u \, du = -\cot u$

8. $\displaystyle\int \sec u \tan u \, du = \sec u$

9. $\displaystyle\int \csc u \cot u \, du = -\csc u$

10. $\displaystyle\int \tan u \, du = -\log |\cos u|$

11. $\int \cot u \, du = \log |\sin u|$

12. $\int \sec u \, du = \log |\sec u + \tan u|$

13. $\int \csc u \, du = \log |\csc u - \cot u|$

14. $\int \dfrac{du}{u^2 + a^2} = \dfrac{1}{a} \tan^{-1} \dfrac{u}{a}$

15. $\int \dfrac{du}{u^2 - a^2} = \dfrac{1}{2a} \log \left| \dfrac{u-a}{u+a} \right|, \, u > a$

16. $\int \dfrac{du}{a^2 - u^2} = \dfrac{1}{2a} \log \left| \dfrac{a+u}{a-u} \right|, \, u < a$

17. $\int \dfrac{du}{\sqrt{a^2 - u^2}} = \sin^{-1} \dfrac{u}{a}$

18. $\int \dfrac{du}{\sqrt{u^2 \pm a^2}} = \log |u + \sqrt{u^2 \pm a^2}|$

19. $\int u \, dv = uv - \int v \, du$

20. $n \int \cos^n x \, dx = \cos^{n-1} x \sin x +$
$(n-1) \int \cos^{n-2} x \, dx$

21. $n \int \sin^n x \, dx = -\sin^{n-1} x \cos x +$
$(n-1) \int \sin^{n-2} x \, dx$

22. $\int \dfrac{du}{a + bu} = \dfrac{1}{b} \log |a + bu|$

23. $\int \dfrac{du}{(a + bu)^2} = -\dfrac{1}{b(a + bu)}$

24. $\int \dfrac{du}{(a + bu)^3} = -\dfrac{1}{2b(a + bu)^2}$

25. $\int \dfrac{u \, du}{a + bu} = \dfrac{1}{b^2}(a + bu - a \log |a + bu|)$

26. $\int \dfrac{u \, du}{(a + bu)^2} = \dfrac{1}{b^2}\left(\log |a + bu| + \dfrac{a}{a + bu}\right)$

27. $\int \dfrac{u \, du}{(a + bu)^3} = \dfrac{1}{b^2}\left(-\dfrac{1}{a + bu} + \dfrac{a}{2(a + bu)^2}\right)$

28. $\int \dfrac{u^2 \, du}{a + bu} =$
$\dfrac{1}{b^3}\left[\tfrac{1}{2}(a + bu)^2 - 2a(a + bu) + a^2 \log |a + bu|\right]$

29. $\int \dfrac{u^2 \, du}{(a + bu)^2} =$
$\dfrac{1}{b^3}\left[a + bu - 2a \log |a + bu| - \dfrac{a^2}{a + bu}\right]$

30. $\int \dfrac{u^2 \, du}{(a + bu)^3} =$
$\dfrac{1}{b^3}\left[\log |a + bu| + \dfrac{2a}{a + bu} - \dfrac{a^2}{2(a + bu)^2}\right]$

31. $\int \dfrac{du}{u(a + bu)} = -\dfrac{1}{a} \log \left| \dfrac{a + bu}{u} \right|$

32. $\int \dfrac{du}{u^2(a + bu)} = -\dfrac{1}{au} + \dfrac{b}{a^2} \log \left| \dfrac{a + bu}{u} \right|$

33. $\int \dfrac{du}{u(a + bu)^2} =$
$\dfrac{1}{a(a + bu)} - \dfrac{1}{a^2} \log \left| \dfrac{a + bu}{u} \right|$

34. $\int \dfrac{du}{u^2(a + bu)^2} =$
$-\dfrac{a + 2bu}{a^2 u(a + bu)} + \dfrac{2b}{a^3} \log \left| \dfrac{a + bu}{u} \right|$

35. $\int \sqrt{a + bu} \, du = \dfrac{2}{3b} \sqrt{(a + bu)^3}$

36. $\int u \sqrt{a + bu} \, du =$
$-\dfrac{2(2a - 3bu)\sqrt{(a + bu)^3}}{15b^2}$

37. $\int u^2 \sqrt{a + bu} \, du =$
$\dfrac{2(8a^2 - 12abu + 15b^2 u^2)\sqrt{(a + bu)^3}}{105b^3}$

38. $\int \dfrac{du}{\sqrt{a + bu}} = \dfrac{2\sqrt{a + bu}}{b}$

39. $\int \dfrac{u \, du}{\sqrt{a + bu}} = -\dfrac{2(2a - bu)}{3b^2}\sqrt{a + bu}$

40. $\int \dfrac{u^2 \, du}{\sqrt{a + bu}} =$
$\dfrac{2(8a^2 - 4abu + 3b^2 u^2)}{15b^3}\sqrt{a + bu}$

41. $\int \dfrac{du}{u\sqrt{a + bu}} =$
$\dfrac{1}{\sqrt{a}} \log \left| \dfrac{\sqrt{a + bu} - \sqrt{a}}{\sqrt{a + bu} + \sqrt{a}} \right|, \text{ if } a > 0$

42. $\int \dfrac{du}{u\sqrt{a+bu}} = \dfrac{2}{\sqrt{-a}}\tan^{-1}\sqrt{\dfrac{a+bu}{-a}}$, if $a < 0$

43. $\int \sqrt{a^2 - u^2}\, du = \dfrac{1}{2}\left(u\sqrt{a^2 - u^2} + a^2 \sin^{-1}\dfrac{u}{a}\right)$

44. $\int u\sqrt{a^2 - u^2}\, du = -\dfrac{1}{3}(a^2 - u^2)^{3/2}$

45. $\int u^2\sqrt{a^2 - u^2}\, du = \dfrac{u}{8}(2u^2 - a^2)\sqrt{a^2 - u^2} + \dfrac{a^4}{8}\sin^{-1}\dfrac{u}{a}$

46. $\int \dfrac{\sqrt{a^2 - u^2}}{u}\, du = \sqrt{a^2 - u^2} - a\log\left|\dfrac{a + \sqrt{a^2 - u^2}}{u}\right|$

47. $\int \dfrac{u\, du}{\sqrt{a^2 - u^2}} = -\sqrt{a^2 - u^2}$

48. $\int \dfrac{u^2\, du}{\sqrt{a^2 - u^2}} = -\dfrac{u}{2}\sqrt{a^2 - u^2} + \dfrac{a^2}{2}\sin^{-1}\dfrac{u}{a}$

49. $\int \dfrac{du}{u\sqrt{a^2 - u^2}} = \dfrac{1}{a}\log\left|\dfrac{a - \sqrt{a^2 - u^2}}{u}\right|$

50. $\int \dfrac{du}{u^2\sqrt{a^2 - u^2}} = -\dfrac{\sqrt{a^2 - u^2}}{a^2 u}$

51. $\int (a^2 - u^2)^{3/2}\, du = \dfrac{u}{8}(5a^2 - 2u^2)\sqrt{a^2 - u^2} + \dfrac{3a^4}{8}\sin^{-1}\dfrac{u}{a}$

52. $\int \dfrac{du}{(a^2 - u^2)^{3/2}} = \dfrac{u}{a^2\sqrt{a^2 - u^2}}$

53. $\int \dfrac{u\, du}{(a^2 - u^2)^{3/2}} = \dfrac{1}{\sqrt{a^2 - u^2}}$

54. $\int \dfrac{u^2\, du}{(a^2 - u^2)^{3/2}} = \dfrac{u}{\sqrt{a^2 - u^2}} - \sin^{-1}\dfrac{u}{a}$

55. $\int \sqrt{u^2 \pm a^2}\, du = \dfrac{1}{2}\left[u\sqrt{u^2 \pm a^2} \pm a^2\log\left|u + \sqrt{u^2 \pm a^2}\right|\right]$

56. $\int u\sqrt{u^2 \pm a^2}\, du = \dfrac{1}{3}(u^2 \pm a^2)^{3/2}$

57. $\int u^2\sqrt{u^2 \pm a^2}\, du = \dfrac{u}{8}(2u^2 \pm a^2)\sqrt{u^2 \pm a^2} - \dfrac{a^4}{8}\log\left|u + \sqrt{u^2 \pm a^2}\right|$

58. $\int \dfrac{\sqrt{u^2 - a^2}}{u}\, du = \sqrt{u^2 - a^2} - a\sec^{-1}\dfrac{u}{a}$

59. $\int \dfrac{\sqrt{u^2 + a^2}}{u}\, du = \sqrt{u^2 + a^2} - a\log\left|\dfrac{a + \sqrt{u^2 + a^2}}{u}\right|$

60. $\int \dfrac{u\, du}{\sqrt{u^2 \pm a^2}}\, du = \sqrt{u^2 \pm a^2}$

61. $\int \dfrac{u^2\, du}{\sqrt{u^2 \pm a^2}} = \dfrac{1}{2}\left(u\sqrt{u^2 \pm a^2} \mp a^2\log\left|u + \sqrt{u^2 \pm a^2}\right|\right)$

62. $\displaystyle\int \frac{du}{u\sqrt{u^2 + a^2}} = \frac{1}{a} \log \left| \frac{u}{a + \sqrt{u^2 + a^2}} \right|$

63. $\displaystyle\int \frac{du}{u\sqrt{u^2 - a^2}} = \frac{1}{a} \sec^{-1} \frac{u}{a}$

64. $\displaystyle\int (u^2 \pm a^2)^{\frac{3}{2}} \, du = \frac{u}{8} (2u^2 \pm 5a^2)\sqrt{u^2 \pm a^2} + \frac{3a^4}{8} \log | u + \sqrt{u^2 \pm a^2}|$

65. $\displaystyle\int \frac{du}{(u^2 \pm a^2)^{\frac{3}{2}}} = \frac{\pm u}{a^2\sqrt{u^2 \pm a^2}}$

66. $\displaystyle\int \frac{u \, du}{(u^2 \pm a^2)^{\frac{3}{2}}} = \frac{-1}{\sqrt{u^2 \pm a^2}}$

67. $\displaystyle\int \frac{u^2 \, du}{(u^2 \pm a^2)^{\frac{3}{2}}} = - \frac{u}{\sqrt{u^2 \pm a^2}} + \log | u + \sqrt{u^2 \pm a^2}|$

68. $\displaystyle\int \frac{\sqrt{u^2 \pm a^2}}{u^2} \, du = - \frac{\sqrt{u^2 \pm a^2}}{u} + \log | u + \sqrt{u^2 \pm a^2}|$

69. $\displaystyle\int \frac{du}{u^2\sqrt{u^2 \pm a^2}} = \mp \frac{\sqrt{u^2 \pm a^2}}{a^2 u}$

70. $\displaystyle\int \frac{du}{au^2 + bu + c} = \frac{1}{\sqrt{b^2 - 4ac}} \log \left| \frac{2au + b - \sqrt{b^2 - 4ac}}{2au + b + \sqrt{b^2 - 4ac}} \right|$, if $b^2 - 4ac > 0$

71. $\displaystyle\int \frac{du}{au^2 + bu + c} = \frac{2}{\sqrt{4ac - b^2}} \tan^{-1} \frac{2au + b}{\sqrt{4ac - b^2}}$, if $b^2 - 4ac < 0$

72. $\displaystyle\int \frac{du}{\sqrt{au^2 + bu + c}} = \frac{1}{\sqrt{a}} \log \left| \sqrt{au^2 + bu + c} + \frac{2au + b}{2\sqrt{a}} \right|$, if $a > 0$

73. $\displaystyle\int \frac{du}{\sqrt{au^2 + bu + c}} = \frac{1}{\sqrt{-a}} \sin^{-1} \left(\frac{-2au - b}{\sqrt{b^2 - 4ac}} \right)$, if $a < 0$

74. $\displaystyle\int \frac{u \, du}{\sqrt{au^2 + bu + c}} = \frac{\sqrt{au^2 + bu + c}}{a} - \frac{b}{2a} \int \frac{du}{\sqrt{au^2 + bu + c}}$

75. $\displaystyle\int \sqrt{au^2 + bu + c} \, du$
$\displaystyle = \frac{(2au + b)\sqrt{au^2 + bu + c}}{4a} - \frac{b^2 - 4ac}{8a} \int \frac{du}{\sqrt{au^2 + bu + c}}$

76. $\displaystyle\int \sin mu \sin nu \, du = - \frac{\sin (m + n)u}{2(m + n)} + \frac{\sin (m - n)u}{2(m - n)}$, $m \neq \pm n$

77. $\displaystyle\int \cos mu \cos nu \, du = \frac{\sin (m + n)u}{2(m + n)} + \frac{\sin (m - n)u}{2(m - n)}$, $m \neq \pm n$

78. $\displaystyle\int \sin mu \cos nu \, du = - \frac{\cos (m + n)u}{2(m + n)} - \frac{\cos (m - n)u}{2(m - n)}$, $m \neq \pm n$

79. $\displaystyle\int \sin^{-1} u \, du = u \sin^{-1} u + \sqrt{1 - u^2}$

80. $\displaystyle\int \cos^{-1} u \, du = u \cos^{-1} u - \sqrt{1 - u^2}$

81. $\displaystyle\int \tan^{-1} u \, du = u \tan^{-1} u - \tfrac{1}{2} \log (1 + u^2)$

82. $\displaystyle\int \cot^{-1} u \, du = u \cot^{-1} u + \tfrac{1}{2} \log (1 + u^2)$

83. $\displaystyle\int \sec^{-1} u \, du = u \sec^{-1} u - \log|u + \sqrt{u^2 - 1}|$

84. $\displaystyle\int \csc^{-1} u \, du = u \csc^{-1} u + \log|u + \sqrt{u^2 - 1}|$

85. $\displaystyle\int \frac{du}{a + b \cos u} = \frac{2}{\sqrt{a^2 - b^2}} \tan^{-1} \frac{\sqrt{a^2 - b^2} \tan \dfrac{u}{2}}{a + b}, \text{ if } a^2 > b^2$

86. $\displaystyle\int \frac{du}{a + b \cos u} = \frac{1}{\sqrt{b^2 - a^2}} \log \left| \frac{a + b + \sqrt{b^2 - a^2} \tan \dfrac{u}{2}}{a + b - \sqrt{b^2 - a^2} \tan \dfrac{u}{2}} \right|, \text{ if } a^2 < b^2$

87. $\displaystyle\int \frac{du}{a + b \sin u} = \frac{2}{\sqrt{a^2 - b^2}} \tan^{-1} \left[\frac{a \tan \dfrac{u}{2} + b}{\sqrt{a^2 - b^2}} \right] \text{ if } a^2 > b^2$

88. $\displaystyle\int \frac{du}{a + b \sin u} = \frac{1}{\sqrt{b^2 - a^2}} \log \left| \frac{a \tan \dfrac{u}{2} + b - \sqrt{b^2 - a^2}}{a \tan \dfrac{u}{2} + b + \sqrt{b^2 - a^2}} \right|, \text{ if } a^2 < b^2$

89. $\displaystyle\int \frac{du}{a^2 \cos^2 u + b^2 \sin^2 u} = \frac{1}{ab} \tan^{-1} \frac{b \tan u}{a}$

90. $\displaystyle\int \frac{du}{a + be^{nu}} = \frac{1}{an} (nu - \log|a + be^{nu}|)$

91. $\displaystyle\int \frac{du}{ae^{nu} + be^{-nu}} = \frac{1}{n\sqrt{ab}} \tan^{-1} \left(e^{nu} \sqrt{\frac{a}{b}} \right)$

92. $\displaystyle\int e^{au} \sin bu \, du = \frac{e^{au}(a \sin bu - b \cos bu)}{a^2 + b^2}$

93. $\displaystyle\int e^{au} \cos bu \, du = \frac{e^{au}(b \sin bu + a \cos bu)}{a^2 + b^2}$

94. $\displaystyle\int_0^{\pi/2} \cos^n u \, du = \int_0^{\pi/2} \sin^n u \, du$

$$= \frac{(n-1)(n-3)(n-5) \cdots 1}{n(n-2)(n-4) \cdots 2} \cdot \frac{\pi}{2}, \text{ if } n \text{ is even. } > 0$$

$$= \frac{(n-1)(n-3)(n-5) \cdots 2}{n(n-2)(n-4) \cdots 3}, \text{ if } n \text{ is odd, } > 1$$

INDEX

(by Topic and Section)